The Memory We Could Be provides a gripping review of where we've been, where we are, and where we may be headed. Which future will we choose? Will we head down a path of continued environmental degradation rendering the planet unlivable for future generations, or will we act in time to avert catastrophic climate change and environmental ruin? This book makes an impassioned plea that we choose that latter path, and in so doing, assure that that be our legacy, the memory that future generations will have of us.

— Michael Mann, Distinguished Professor, Penn State University and co-author, *The Madhouse Effect*

As we navigate our way through the Anthropocene, we need young writers' voices more than ever. Clear, poetic, and full of insight, Macmillen Voskoboynik's book offers an exhilarating introduction to our ecological crisis, what caused it, and how we can imagine a better future.

— Jason Hickel, anthropologist, Goldsmiths University and author, *The Divide: A Brief Guide to Global Inequality and its Solutions*

This book is a timely call to action to prevent climate breakdown. Those who care about protecting this planet should read Daniel's work and prepare to build a new way of living.

— Caroline Lucas, Co-Leader, Green Party of England and Wales

What is missing in climate change literature are bold, compelling voices, accounts that are accountable to the dignity of the afflicted... Macmillen Voskoboynik's work is a beacon in this regard.

— Asad Rehman, Executive Director, War on Want

Macmillen Voskoboynik offers a sweeping overview of the ecological predicaments and choices that confront us in the 21st century. He's a hopeful realist—exactly the sort of storyteller and analyst we need at this fraught moment.

— Richard Heinberg, author, *The End of Growth*

THE MEMORY WE COULD BE

THE MEMORY WE COULD BE

OVERCOMING FEAR TO CREATE OUR ECOLOGICAL FUTURE

DANIEL MACMILLEN VOSKOBOYNIK

new society
PUBLISHERS

First published by New Internationalist, newint.org

Cover design and digital composite: Diane McIntosh.
Cover images © iStock.
Interior images: p 13 © fottoo; p 107 © MISHELA;
p 165 © diversepixel / Adobe Stock.

Printed in Canada. First printing September 2018.

This book is intended to be educational and informative.
It is not intended to serve as a guide. The author and publisher disclaim
all responsibility for any liability, loss or risk that may be associated
with the application of any of the contents of this book.

Inquiries regarding requests to reprint all or part of *The Memory We Could Be*
should be addressed to New Society Publishers at the address below. To order directly from the publishers, please call toll-free (North America) 1-800-567-6772,
or order online at www.newsociety.com

Any other inquiries can be directed by mail to:

New Society Publishers
P.O. Box 189, Gabriola Island, BC V0R 1X0, Canada
(250) 247-9737

LIBRARY AND ARCHIVES CANADA CATALOGUING IN PUBLICATION

Macmillen Voskoboynik, Daniel, author
The memory we could be : overcoming fear to create our ecological
future / Daniel Macmillen Voskoboynik.

Includes bibliographical references and index.
Issued in print and electronic formats.
ISBN 978-0-86571-899-9 (softcover).—ISBN 978-1-55092-692-7 (PDF).—
ISBN 978-1-77142-288-8 (EPUB)

1. Human ecology—Psychological aspects. I. Title.

BF353.5.N37V67 2018   155.9'1   C2018-903587-0
                                              C201 8-903588-9

Funded by the   Financé par le
Government   gouvernement
of Canada   du Canada

Canada

New Society Publishers' mission is to publish books that contribute
in fundamental ways to building an ecologically sustainable
and just society, and to do so with the least possible impact on
the environment, in a manner that models this vision.

new society
PUBLISHERS

Certified
B
Corporation

MIX
Paper from
responsible sources
FSC® C016245

Contents

PRESENT

Foreword

by RAOUL MARTINEZ

A trail of destruction lies behind our civilization. In its wake are lost languages and cultures, broken bodies and ecosystems. To retrace these steps is to glimpse an inconvenient truth: our society has been built on violence. It impoverished as it enriched, and unleashed chaos as it imposed order. The global distribution of power that we see today was born of slavery, colonialism, theft and war. The riches amassed are inextricable from the plunder and pollution of Earth's oceans, rivers, forests and soil. Through a strange alchemy, the finite wealth of nature has been destroyed in the production of abstract financial wealth. The value of ancient lands and human labor has been transformed into digits on screens that measure the fortunes of the privileged. The result is that 1 percent of the humans who woke up this morning own as much wealth as the other 99 percent combined.

Looking ahead, one thing is clear: the path we are on is coming to an end. Projections of the coming decades—whether they focus on food supplies, conflicts or weather patterns—read like dystopian fiction.

The world we create as individuals, communities and nations is a mirror in which we can glimpse something of ourselves. If a darker future is to be averted, more of us need to join the dots that link ideas to outcomes and values to violence. We need to make profound changes to our thoughts and behavior—changes that will cascade upward toward a transformation of the global systems that dominate our lives. For this transformation to occur, our interdependence needs to be widely recognized. The many boundaries that divide us—psychological and physical—must be transcended. The emotional distance maintained by borders and bank balances, identities and iPhones needs to close. The crises we face demand that competition make way for cooperation, and isolation open up to connection. They demand that the stories we tell about ourselves—economic, cultural, political—be woven into the ever more intricate stories of nature, of which we are but a small part.

The Memory We Could Be is a remarkable contribution to this urgent project. The clarity of its ideas, the depth of its insights and the beauty of its language challenge and inspire. At its heart is a profound sensitivity to the suffering and wisdom of those whose voices are too often ignored. Daniel Macmillen Voskoboynik leads us beyond simple narratives and cold statistics to a nuanced, holistic understanding of the crises we face and the possible futures that lie ahead. I truly hope its message is heard and acted upon.

RAOUL MARTINEZ is the author of *Creating Freedom: Power, Control and the Fight for Our Future*.

CHAPTER 1

THE MIGHT OF MEMORY

We are born children of the earth.
U'WA WERJAIN SHITA traditional authorities

. . . .

To each epoch its own words.
KAZAKH proverb

. . . .

Water is taught by thirst.
EMILY DICKINSON

. . . .

WE ARE MADE OF MEMORY. In our mother's womb, cells weave replicas of our parents' bodies: a heart, a brain, a knotwork of veins, a shelter of skin. The first breath we take, like those that will follow it, pulls particles of past into our chest: strands of oxygen and carbon that have traveled through the lungs and leaves of centuries.

We are born into a universe we will belong to, into a planet formed by billions of years, into an ancestry drawn by generations. Fortune finds us our family. As our umbilical cord is cut, we are bound to less visible cords that tie us to cultures, traditions and societies.

We start the uncertain journey of life. We grow. Our biological memory, the encoded stories of our genes, unfurls. We begin walking, on an Earth that holds the remains of our ancestors.

We acquire names for the world around us. We attempt to express our internal world through language, a memory of words and grammar.

1

With time, we build identities with the mortar of our childhood memories. We interact with fellow humans, who share virtually all of our genetic memory, exploring and exhibiting the remaining fraction that makes us who we are.

Rituals, songs and conversations subtly hand us the lessons of yesterday. The gaze we hold, the dreams we dream and the opinions we form are shaped by our surroundings.

As we age, new selves graft over the old. Memory, the tide of re-membering and forgetting, retains and releases, defining us. And so we live our lives, carrying our unprecedented story, striving to write the memory we would like to be.

▪ ▪ ▪ ▪ ▪

Nature is the memory of the Earth. Behind every forest, every val-ley, every body of water is a hidden history, a patient effort of time. Landscapes are carved by wind and water. Trees and plants are sculpted by the hands of altitude, precipitation and sunlight.

Time flows, and in its currents, existence leaves its mark. Trees etch rings, faces trace wrinkles, sediments fold layers and whales archive journeys in their baleen.[1]

Life passes itself on. Like seeds, our own societies disperse their memory. Farmers rely on our memory of agronomy to nourish life from soil. Educators and elders transmit human memory through stories, told and written. Lawyers interpret and apply juridical memory. Doctors examine patients, drawing lessons from our his-tory of healing.

Scholars devote lakes of ink to documenting and explaining our evolving memory. Historians reconstruct the past, the imprint of human endeavor. Astronomers turn their telescopes to the skies, watching the delayed memory of stars arrive from light years away. Geologists trace the movement of mountains over millennia. Ar-chaeologists brush away the sediments of time. The present re-leases memory, and journalists rush to record the latest events.

Together, in our own ways, we recall and rewrite the memory of human survival.

▪ ▪ ▪ ▪ ▪

We all are because others are. Born of love, we begin as delicate beings in need of care. Our parents, our grandparents and our communities are the immediate forces that bring us into this world. Yet we are also the descendants of unknown predecessors, both human and nonhuman, that have come before us.

The Earth is only habitable for humans because of the minuscule organisms that breathed oxygen into our atmosphere millions of years ago. Our own life form today is the result of a persistent transition from cells into bacteria, from organisms into diverse species. The elements that compose us originate in the stars.

We are small strokes on the vast canvas of time. The Earth that sustains us is over four and a half billion years old. In comparison, our life as a species begins only 200,000 years ago. If the history of our planet were to be made into a two-hour film, human beings would only feature in the final second.[2] But that final fragile second holds an infinite sea of stories. Stories of loves and longings, of joys and sadnesses, of wishes and wonders. Stories about the creation and protection of life, and stories of its eradication.

⬚   ⬚   ⬚   ⬚   ⬚

Human beings cannot live without forgetting. Inhibiting memory is a bodily function. But, unlike our minds, the wider world does not forget.

The living memory of our planet narrates its state. Wherever we look, every sphere of life—our atmosphere, our biosphere, our hydrosphere, our lithosphere, our cryosphere—is marked by destruction.

Over recent centuries, a portion of humanity has radically disrupted the cycles of the planet's waters, soils and thermal balance. As a result, we have entered an age of ending, where we are extinguishing the conditions necessary for our own survival.

We are dismantling our own existential stage, setting in train a slow-motion genocide where crimes against humanity are obscured by their frequency. In doing so, we are wrecking our human heritage, shredding safety nets and condemning the world's most vulnerable to fates that defy transcription.

Seas are heating, rising and acidifying. Poles of ice are melting, experiencing the highest rates of warming on Earth. Ice sheets are increasingly losing mass. Glaciers, the water towers of valleys, are retreating. Our oceans, which hold most of the Earth's living space, are an exhibition of extinctions. The Great Barrier Reef, the planet's largest living structure, is in terminal decline, disintegrating in warming waters. Entire marine ecosystems are disappearing, with 90 percent of the world's fisheries collapsing or fully exploited.[3]

Our forests have been razed and despoiled. Half of the global tree population has been decimated, with 15 billion trees cut annually.[4] Entire mountains have been gutted, as mines stretch deeper scars into the Earth. Our intensive systems of agriculture have similarly carved exhaustion into the roots of land.

We are simply removing life faster than it creates itself. Every year, our relentless withdrawal of natural resources significantly surpasses the Earth's ability to regenerate those resources. Our degradation and erosion of lands overtake their ability to form and replenish fertile soil. Our production of waste outpaces the planet's ability to safely absorb that waste. We are overdrafting aquifers, overgrazing pastures, overcultivating soils and overloading our atmosphere.[5]

The pace of extraction shows little sign of ebbing. Instead, we are accelerating trends, hastening fertilizer consumption, water use, forest clearance and marine animal capture.

Our world's wealth is its diversity, but our assault on our own home is driving widespread extinction. The imposition of devastating development models has laid waste to thousands of cultures and ways of being. A human language becomes extinct at the rate of one per fortnight.[6] Cornered by deforestation, pollution and poaching, 100 species are being lost a day. The global population of fish, birds, mammals, reptiles and amphibians declined by 58 percent between 1970 and 2012.[7] In the next 30 years, 90 percent of all marine species may have been lost. By that time, there could be more plastic in our oceans than fish.[8] Current rates of species extinction are hundreds of times greater than the geological norm, implying what scientists have called a "frightening assault on the foundations of human civilization."[9]

Pollution has hidden the stars, poisoned our waters and ravished the lungs of our children. It has turned communities into cancer villages, city residents into smog refugees and billions of people into the living proof of a sanitary emergency. Around 92 percent of the world's population is exposed to levels of air pollution above the World Health Organization's guideline levels.[10] Half of New Delhi's schoolchildren have permanent lung damage.[11] In 2015, pollution was responsible for over two and a half million deaths in India alone.[12]

Scientists note our breach of planetary boundaries, the key biophysical guardrails which allow for safe human life. We have already crossed, or are set to cross, a whole range of these limits: ocean acidification, biological diversity loss, the disruption of nitrogen and phosphorus cycles, disappearance of fresh water, changes in land cover, growing pollution from synthetic chemicals, ozone depletion, toxic chemical pollution and the loading of atmospheric aerosols.[13]

Today, we are seeing record rates of fossil-fuel burning, accompanied by significant rises in emissions. Our seas, skies, lands and winds are all in flux. Virtually every continent, region and country is indicating record or near-record levels of heat. Fires, ferocious storms, torrential rains and droughts are occurring with increasing frequency and intensity.

Of the hottest 17 years on record, 16 have occurred since 2000. In the past 40 years, the percentage of our planet affected by drought has doubled. Since 1970, the number of extreme weather events occurring every year has quadrupled.[14]

Weather records are being obliterated, but even the notion of a record makes little sense today. Under current trends and scenarios, the "new normal" may be a world where the barrier of expectation is always pushed further back, a horizon of pain in constant retreat.

Most governments, particularly the world's richest, are failing to meet their own meager pledges for action. The majority of global institutions find themselves with their backs to reality.

And the unrelenting pressures our environments are subjected to mirror those that pervade in human societies, defined by acute poverties, inequalities and avoidable deprivations. Through the atmospheric violence we have unleashed, we risk even further reinforcing these injustices. Ours is a world fertile for suffering.

▪ ▪ ▪ ▪ ▪

These sorrowful realities, the silent signatures of centuries, are monuments to a crisis that evades its diagnosis. Climate change, rather than a root problem, is the salient symptom of a human world unwilling or unable to protect its own life.

Our ecological predicament is not an anomaly, a small setback on our treadmill of progress. It is not a mere outcome of an absence of leadership, education or technology.

Rather, it is a civilizational crisis. A crisis of our dominant thinking, which has for too long neglected what is needed to sustain existence, and a crisis of our economic model, which roots development in destruction.

Considering these discomfiting conclusions is not easy. Perhaps the prime impediment to understanding climate change is fear. A fear of pain. A fear of grief. A fear of implication and guilt. A fear of challenging our precious presumptions, and overturning our worldviews. A fear of failure. A fear of losing comforts. A fear of violating the innocent dream that the world will be okay.

These fatal fears pull us toward apathy, toward denial, toward desolation, toward false hope. We avoid the topic. We adjust ourselves to its magnitude, assembling a psychology of faith. We rush impressions and opinions. We question the credibility of evidence, for its inferences are incredible. We nourish beliefs in happy endings, in imminent solutions, in technical fixes, in painless paths to safety. Thinking it is beyond us, we turn to apocalyptic dejection, often another shape of indifference.

And so, forced by fear, the topic of climate change sinks into silence. It gains the status of death, tainted by triviality and taboo.

▪ ▪ ▪ ▪ ▪

The first step to overcoming fear is acknowledging its presence. By admitting our fears, we can start to transcend them and redirect their force. This book hopes to overpower the impotence climate change creates in us.

This fear is compounded by our ecological illiteracy. Our societies rarely, if ever, devote attention to examining the genesis, repercussions or stakes of environmental problems. Disoriented, we struggle to read our surrounding realities, to unpack blaring news stories or find the relevance in intangible alarm. Confusion, when coupled with perceived distance, can easily lend itself to callousness.

This book brings together the insights of many disciplines to try to clarify our ecological reality, and its possible trajectories. It starts by peering into the black box of our human past to understand how we got here.

It then proceeds to illustrate how climate violence is shaping our world today. The framing of environmental danger is often associated with the future. But our ecological crisis is not an abstraction we may hand to our grandchildren. It is not an advertisement, a warning or a hypothesis. It is contemporary, shatteringly and definitively so.

Finally, the book's closing chapters hope to sketch out the future, showing in human terms the world we can lose and the world we can win.

Authority and humility

The world around us is astoundingly complex. Contemplating the infinite intricacy of the natural world, botanist Frank Egler observed that "ecosystems are not only more complex than we think but more complex than we can think."[15]

Reality escapes our simplifying control, defying explanation and measure. Our fragile formulae and delicate strands of insight will always be outweighed by our endless ignorance. The world always knows more than we do.

Climate change is particularly vulnerable to the temptations of certainty. When we talk about planetary problems, it becomes easy to confuse the big picture with the only picture. It becomes tempting to reach for blanket explanations and swift diagnoses: that human nature is to blame; that not enough people know about climate change; that its implication won't be too grave; that human beings will always learn to adapt.

It becomes intuitive to forget who we mean by "we." Although climate change binds our fates as human beings, not everyone is equally responsible for it, and not everyone is equally affected.

It also becomes attractive to seek all the answers in one worldview. Depending on our inclination, we may see particular promise in technology, education or politics. But climate change is a wicked problem, resistant to single solutions, its roots woven into economies, cultures, livelihoods and habits. It traverses every sector of society and every level of human relations. Every perspective, from law to agronomy, medicine to oceanography, is relevant in addressing it.

In this book, I have tried to draw from as many different voices as possible—from anthropology to astrobiology, physics to economics, hydrology to history. I have relied on published scholars, but also on the world's thinkers without diplomas, its scientists without laboratories.

There are additional reasons for assembling this chorus of backgrounds. For too long, the dominant conversation on climate change has included only a tiny range of people, namely a handful of policymakers and valuable scientific sources. This selectivity sidelines the contributions of popular, personal, local and indigenous knowledges, which will be vital if we are to attain any plausible climate safety.[16] To tackle arguably the deepest problem we have ever faced, we are going to need to pull together our collective wisdom, in its plurality of lenses and expressions.

This exclusion not only restricts our gaze, but helps to misrepresent the gravity of the problem. Rarely are the protagonists of pain, those most vulnerable to climate violence, near the spotlight of attention. Without their voices and visions, the story of our environmental reality is evacuated of urgency.

Over decades, the story of climate change has been predominantly encoded in the language of data, diagrams and jargon. It has been poorly illustrated through the narrow iconography of polar bears, collapsing glaciers and stylized temperature graphs. Its relevance has been defused through acronyms, abstract numbers and tired metaphors.

Our own imaginations suffer as a result, struggling to comprehend the emotional density of our unfolding disaster. In this book, I have tried, unsuccessfully, to compile existent narratives that distance themselves from technical and sanitized language. Our words ultimately convey the world. By diluting our vocabulary, we bleach the problem.

Finally, given the intricacy of the topic, I must also admit that the words that follow are written with hesitation. Often books will carry confident conclusions, their neat structures arranging orderly arguments that flow directly from point and point.

These pages, however, carry more questions than certainties. Every phrasing and assertion presented here is tentative. When faced with an impossibly intricate world, and crises that transcend their definition, there are few assured answers or absolutes. This book is just one small attempt to assemble perspectives and instruments that may assist us in the pursuit of greater literacy.

An old teacher of mine compared learning to walking through an endless corridor of veils. With every curtain we peel away, we find ourselves before another. The more we learn, the less we can generalize. In any text, there are inevitable tensions between concision and comprehensiveness. In advance, I apologize for all simplifications made and shortcuts taken.

While reading, it is always worth remembering that there is no single story about climate injustice. Our world is plural, just like its people. There are infinitely many ways to tell a tale; this is just one, the product of my restricted field of vision, ridden with all its inevitable errors. I hope not to impose any totalizing narratives, or oversell any theories.

I mention this not merely as a disclaimer, but as a vital point. As I hope this book will show, if we want to accomplish any degree of climate safety, humility will need to gain prominence in our problem-solving strategies. After all, the word "humility" shares its root with the word "human": of the *humus*, of the earth. To be grounded is to be attentive to our limitations, to be aware that all of us bear forms of myopia we cannot correct. Our horizons are not the edges of reality.

With this in mind, I want to be transparent about my own biases. I will not pretend to talk about climate change from some removed stance of impartiality. Much of how I think is shaped by what is sometimes called climate justice.

Climate justice is a small lens through which we can view the world. It asks us to be guided by justice, by a sensitivity to what is fair. It urges us to be attentive to history, to acknowledge wrongs that have been wrought and consider how to redress them. It encourages us to tackle the problem at its root, and ensure that by tackling climate change, we are building a more beautiful, equitable society.

For all its shortcomings, this is a way of thinking that also holds a few useful insights. Those most affected by climate change are those least likely to have caused it. Those most likely to have caused it are those most likely to avoid its negative effects. The brutality leveled at the nonhuman world runs parallel to the brutality of the human world. The inequalities that have generated our climate crisis are also those that inhibit our response. To overcome our ecological predicament, we need to transform the structures that have generated it.

■   ■   ■   ■   ■

The trends are terrifying, and every day we rush farther into uncharted territory. There are sadnesses we won't be able to avoid. But what seems certain is that human action over the next years will determine whether we will face grave loss or catastrophic collapse. It is our responsibility to equip ourselves for these defining years. Learning to think ecologically is a precious and indispensable tool for our times. As José Lutzenberger noted: "ecology is the science of the symphony of life, it is the science of survival."[17]

The fight to tackle climate change is a fight to determine the fatality of the future. A fight over the vindication of life. It will require much of us: to unlearn our despair and learn our possibilities. But through its rigors, we can win a more beautiful world. We can live and create the wanted memory of tomorrow.

PAST

CHAPTER 2

SEPARATION

A civilization that chooses to close its eyes
to its most crucial problems is a stricken civilization.
AIMÉ CÉSAIRE

◼ ◼ ◼ ◼ ◼ ◼ ◼ ◼ ◼ ◼

The water we ingest every day becomes the plasma of our blood,
the steam we exude, the fluid freed by our tear ducts. Sooner or later,
due to the environmental damage of one [oil] well, or the accumulated
damage of hundreds or thousands of wells, we will be left without
the water we need to sweat, to cry of happiness or sadness....
If [oil & gas extraction] were to start in our province,
we would literally pay for it with blood, sweat and tears.
JAVIER CABROL, student, Entre Ríos, Argentina[1]

◼ ◼ ◼ ◼ ◼ ◼ ◼ ◼ ◼ ◼

WE ARE THE AIR WE BREATHE, the land we walk, the food we feed on and the skies we dream under. We are the rivers that sing in streams and taps, the roads that carry us, the winds that shake us and the stars that guide us. We are the water that cleans us, the places that place us and the dust of our ancestors. We are the thoughts we think, the beings we birth, the colors we see, the bodies we heal, the harvests we reap, the homes we make, the families we inherit, the friends we find and the lives we weave.

No matter where we may be, we are inseparable from nature. We live in nature, we live with nature, we rely on nature, and we are part of nature. Our bodies are nature, feeding on and feeding into the energy of the world. All existence flows from our visible and invisible encounters with ecosystems.

15

Everything we were, are and will be is shaped by nature. Our culture—from our agriculture to our architecture to our cuisine—defines how we live, and is in turn defined by the settings it takes root in. In this stark sense, the effacement of the climate is our own destruction. We are the wreckage we struggle to see.

The ecological crisis is a strange, silent and staggering act of self-destruction. Yet our language hides this. When we use words like "nature," "environment" or "climate," we often paint a deceptive portrait of independence. To speak of environmentalism misleads us, suggesting that the natural world is somehow distinct from the human world, and needs to be protected. Phrases like "natural resources," "natural disaster" or "natural parks" inflect our conversations.

It is hard to rid ourselves of the limiting language that divides us from nature. But this division is at the core of our ecological crisis. The word "crisis" itself stems from the Greek *krisis*, separation.

Our separation from nature is encoded across our societies. Our universities and schools divide subjects into natural sciences and humanities, wedging apart the material and immaterial worlds. Politicians deposit issues into their respective boxes. The environment is relegated to a ministry, as if nature had little to do with health, education or economy. Our newspapers publish pieces on "environmental issues" written by "environmental correspondents." Such distinctions, although arbitrary, disguise themselves as absolute.

Separation opens the gate to cruelty. In order to wreck nature, we must dehumanize it, removing ourselves from it. Destruction requires denigration. The theoretical extraction of human beings from nature, the silencing of our dependencies, is partly what has allowed for the unhinged extraction of nature. We can do no known harm to what we do not need, or what we are not part of.

The idea that humans are simply outside of nature has ancient roots, too extensive for this book's brevity. But suffice it to say, through teachings, beliefs and built blindspots, nature began to be seen in many societies as something exterior. Powerful ideas, in their different incarnations, grounded a separation from nature that would be promoted and pursued for centuries.

Nature was seen as something to be dominated, battled and overcome. Nature was a limitless resource: its trees were timber, its mountains were mines, its peoples were labor, its fields were factories, its animals were livestock, and its seas were fisheries.

Nature was an object, stripped of sentience. Nature was an obstacle, a frontier, to be cleared and conquered. Nature was disorder, in need of subjection to a human order. Nature was fertility, inexhaustible. Nature was productivity, to be optimized and increased. Nature was a code, decipherable. Nature was exoticism, accessible through zoos, parks and pictures. Nature was a mere machine: manipulable, understandable and amenable.

Through such maneuvers of distance, environmental violence became a tradition. Nature, the constitution of life, was considered as little more than a site of plunder. Models of economic development rooted in the exploitation, destruction and abandoning of nature forced themselves into legitimacy.

An arrogance was instilled, allowing certain parts of humanity to place themselves at the center of the world, and conceive themselves as rulers of the Earth.

These ideas rippled across traditions of science and politics, that increasingly sought to simplify, predict and control the world.[2] The French philosopher René Descartes argued that, through reason, we could become "masters and possessors of Nature." The economist Adam Smith regarded nature as "no more than a storehouse of raw materials for man's ingenuity."[3]

Philosophers saw the world before them as knowable and solvable, stitched together by universal laws and rules. All behavior was linear, dictated by clear causes, which produced clear effects. Reality could be distilled into disciplines, broken down into elements that could be modeled and controlled. Drawing on these confidences, thinkers and politicians envisioned a society of perfection: perfect individuals, perfect systems, perfect markets, perfect interventions, perfect ideas. There were no limits to applied human

rationality. Every problem was just a symptom of absent ingenuity. In 1785, the French mathematician Marquis de Condorcet pronounced: "All the errors in politics and in morals are founded upon
philosophical mistakes, which, themselves, are connected with
physical errors."

Any obstacle could be thought out of the way. By constantly re-
fining our experiments, technologies and theories, we could conquer the retreating terrain of unknowing. Progress toward truth
was irrepressible.

And amid these attempts to classify, clarify and control the impossibly intricate world around them, a small group of humans
began to cement powerful distinctions: between matter and mind,
between reasoning and feeling, between purity and impurity, between humans and nonhumans, between us and others, between
men and women, between the normal and abnormal, between the
worthy and the unworthy, between the abled and the disabled, between the enlightened and the ignorant, between modernity and
tradition, between civilization and savagery, between backward-
ness and progress, between human life and its home.

As physicist David Bohm would later warn, "guided by a frag-
mentary self-world view, [humankind] then acts in such a way as to
try to break [itself] and the world up."[4]

Relinquishing a way of thinking

The imprints of separation are found across many of our societies
and philosophies. We have carved hierarchies of humanity, invent-
ing or exaggerating differences between peoples, bodies and traditions.[5] Through force, these fabricated fictions have translated into
firm exclusions and inequalities.

Violence requires the violent to distance themselves from the
victim. To commit violence against human beings, we must strip
them of the humanity we share. We must turn equals into others,
inferior or subhuman. We must convince ourselves that they de-
serve our malice, that their lesser worth merits cruelty, that their

"bad choices" justify the pain of punishment. Our war waged on nature has relied on a similar moral distance.

The severed category of nature has, over history, included much of what was deemed disposable and exploitable: animals, women, slaves, lesser peoples. The supremacy of separation helped to position dispossession as civilization, domination as nurture and oppression as achievement. Those with power easily externalized their problems, offloading destruction onto those unworthy of its absence.

The penchant to separate, coupled with the promise of a controllable world, brought with it the idea that all problems have exact causes and solutions. Today, this dream pervades most professions and ideologies. Depending on our outlook, we might think that enough research, or innovation, or persuasion, or incentives, or deregulation, or thoughtful intervention or revolution can transform our reality. Segregated into our own specialities, we dive into our disciplines to identify potential answers: new theories, new policies, new medicines, new devices, new campaigns, new technologies.

There is no doubt that this specialized push for knowledge over centuries has advanced our thinking in many directions. In many ways, it has served us tremendously well, delivering insights across a range of fields. But its usefulness subsides in the face of today's problems.

How can we find singular solutions to what confronts us today, an overlapping concoction of violence, inequality, hunger, climate change, ecological degradation, racial oppression, corruption, militarism, surveillance, patriarchy and poverty?

Beginning to understand the depths of our problems means ridding ourselves of the illusion of control, and stopping before forcing the disarray of reality into neat boxes.

This is not easy, given the strength of separation's spell. Once we are accustomed to mentalities, it is hard to escape them. Our upbringing and education have trained us from an early age to cut the world into pieces. We have been taught to unsee relationships, and impose clarity onto a world that eludes it. Our gaze instinctively reaches for categories. With minds rigidly adjusted to simplicity, approaching complexity can be frightening. Reassuring faith in perfect answers allows us to escape the sad reality: that things are complicated, that there can be multiple legitimate claims on truth, and that decisions often carry trade-offs.

Making connections

Ecology is the study of the opposite of separation: connection. It is concerned with the relationships and interactions of life, the logic of links that underpins the world. These cannot be demarcated or separated. As biologist Barry Commoner noted, the first law of ecology is: "Everything is connected to everything else."[6]

Look and feel around you. At the ground beneath your feet, the air in your lungs, the pulsing blood of your body, the sky before your eyes, the buildings ahead. Life, escaping boxes and boundaries, is bursting, in a dance of ongoing physical, biological, chemical, climatological and geological processes.

To put on the glasses of ecology is to see relatedness and co-existence. Ours is a world of multilayered interactions, cycles, flows and dependencies. Every organism is a collective, a community of beings. Virtually all animals and plants hold countless life-forms that live inside and aside them. What may look like a single tree is an ecosystem of fungus, bacteria, protists and nematodes. Every root or leaf is a dazzling network of species.[7]

Our own bodies are microbiomes, hosting multitudes of microbes that influence our bodies through relations of symbiosis. They train our immune systems, protect us against disease and even shape our organs. The majority of our own cells are not human, but fungal or microbial.[8]

Even what may appear to be purportedly inanimate, whether it be a boulder or a built home, is in fact, a vibrant assembly of atoms. The world, as theologian Thomas Berry noted, is not a "collection of objects," but a dense "communion of subjects."[9]

Ecology looks at the embroidery of life, and the rich intertwining of its threads. The world is dynamic, constantly shifting, an ongoing conversation between the past, present and future.[10] Nothing is independent, only interdependent, absorbed in relations. Everything attends to each other, shaping and being shaped by the other.[11]

Corals, the forests of the seas, have neurons tuned to moonlight. Trees draw on air to build themselves and then feed the air they used. The waters that pulse through rivers, begin in mountains, traverse valleys and end in seas, before returning to their origin as rains, the lakes of the sky. The soil beneath our feet is a raucous process of creation. Plants grow feeding on the sun, releasing sugars into soils, feeding bacteria and microbes. Similarly, as human beings, we depend on ecosystems for our survival, yet the health of those ecosystems depends on how we treat them as we survive.[12]

Constant connection and co-evolution come hand in hand with the other bedrock of ecology: diversity. The more we look at the world around us, the more we realize an exuberance that cannot be surmised. No form of life is authentic or identical to one another. Each of us are unique beings, the inimitable temporary results of innumerable relations and conditions.[13]

Such constant deviation blunts our ability to generalize. Although our textbooks may tidily arrange nature into kingdoms, phyla, genera and species, ultimately life and nature confound labels and classifications.

From machines to organisms

Dominant understandings of science, economics and philosophy are guided by mechanistic ideas. They view the world as a clock-like machine, made up of parts that can be taken apart, analysed, fixed and reassembled. Ecological thinking strives, by contrast, to be organic: to see the world as a living organism rather than as an inanimate machine. These organisms are made up not of parts, but of participants. These participants, rather than replaceable parts, are components with intricate identities.[14]

To think ecologically requires shedding a lot of the conceptual baggage inherited across centuries. It means viewing a simplified world as a complex world, shaped by realities that are irreducible. It means shifting our gaze from objects to relationships, from isolations to systems, from independent to interactive variables, from binaries to multiplicities, from single reasons to multiple explanations, from linear to dynamic equations, from essences to processes, from total to bounded rationalities, from autonomous individuals to relational human beings, from separate parts to integrated wholes.

Through these transitions, the logics of separation, simplicity and uniformity are replaced by an attentiveness to connection, complexity and diversity. The attempt to achieve certitude and eradicate all uncertainties ceases as we work to acknowledge and manage uncertainty. The pursuit of absolute truth ends as we accept the possibility of various truths, rooted in different values and visions. Humans lose their position at the center of the universe, their role as dictators over nature, to realize that we are dictated by our interrelationship with nature. The urges of domination and control are replaced by ethics of care.

Such comprehensive shifts of thinking are as significant, and as challenging, as the Copernican turn. Yet across the sciences and humanities, from discipline to discipline, new findings are gradually taking us in this direction. Systems of life across the universe are increasingly being seen as staggeringly complex, chaotic and interdependent. In our new paradigm, the world reveals itself to be an endlessly varying, self-organizing place, immersed in irreducible intentions.

Already, the ways in which we approach life are shifting, upending previous schemas and dissolving established boundaries. Breakthroughs in nonlinear mathematics have shed light on indeterminacy and chaos theory. In physics, quantum mechanics has revealed the inadequacy of linear laws of motion. Absolute conceptions of time and space lose their cogency within understandings of relativity.

New pathways in molecular biology and psychology show the intertwined relationship between nature and nurture. Our behavior is shaped by the dense interaction of predispositions, but the expressions of these predispositions depend on their contexts. The conventional view of heredity drew a firm distinction between the "genotype" and the "phenotype." But empirical research and the growing field of epigenetics have shown how environments and experiences can shape genetic inheritance. Genes do not simply become phenotypes but are transformed through multiple processes. Heredity is beset by variation.[15]

Evolution, rather than a linear movement with clear junctures, is viewed as a messier process, where shifts occur at different rates. In zoology, organisms are increasingly not being seen as standalone units or single species, but as networks or symbioses.[16]

In neurology, we have moved from conceiving the brain as split into regions, but stitched together by networks. Rather than a mind removed from matter, new understandings of cognition see the mind as interwoven with matter across the web of life.[17]

Contemporary scientists are facing the immense challenge of finding clarity amid a torrent of new insights that challenge ingrained assumptions and frames.[18]

These insights of intricacy are shattering previous confidences, namely our faith in our ability to neatly order the world. Our belief in our capacity to mathematize, quantify and plot reality. Prediction is fraught with uncertainty. The answers to most research questions become, well, it's hard to tell.

Exposing the fragility of our knowledge is uncomfortable for many of us. The charm of precision, or determinism, has enormous appeal. Acknowledging complexity yields major challenges to orthodoxies, whether they be in jurisprudence, medicine or economics.[19]

Looking forward

To acknowledge complexity is not to discard all previous thought. There are good reasons for simplifying, not least because we separate and mark categories to aid our memory. As primatologist Robert Sapolsky notes: "Putting facts into nice cleanly demarcated buckets of explanation... can help you remember those facts better. But it can wreak havoc on your ability to *think* about those facts."[20]

Complexity urges us to contemplate how we think about the world. When we observe any phenomenon, we impose time, space and language on reality. Anything can be looked at through a different order of magnitude. Reality looks different, depending on whether it is seen through a keyhole, a naked eye, a telescope or a microscope.[21]

Every study, or attempt to depict the world, bears similar assumptions. All models respond to chosen methodologies. All algorithms and equations flow from their internal expectations. All data is defined by its quality. All rankings depend on the weight of their indicators. Any statistics are shaped by their formulation. All averages will hide immense individual variabilities. Every study will provide different findings, depending on the time frames, dimensions, instruments and scales of analysis employed.

But all assumptions can be easily disputed. Rather than seeing theories and models as replicas of reality, we should see them as imperfect artifacts designed to aid our understanding of reality. To observe the world, no matter how useful our insights and instruments may be, is to flash a headlight on a sea of darkness.

These limitations do not invalidate or discredit science, but they do expose its fragility. Reality is messy. Virtually every phenomenon we see is multifactorial and multicausal. Causation is immensely tricky to determine among infinite variables. Every crisis is buried within several others. Rarely are there definitive truths, for knowledge is provisional. As time passes, older theories give way to even more intricate insights, destabilizing previous paradigms. The laws of Newtonian physics, deemed irrefutable in their wake, were overwhelmed in the 20th century.

Ecological thinking does not aim to comprehend everything. It reminds itself of its limits, its insufficiency before the intricacy of the world around us. Instead of aiming to reveal universal laws, it asks us to look at contexts. Some situations will merit simpler approaches. But other situations will not lend themselves to absolute answers, where the normal scientific method can be easily applied. In such post-normal contexts, as philosophers Jerome Ravetz and Silvio Funtowicz note, where facts are uncertain, ethical values are disputed, stakes are high and decisions are urgent—new approaches are needed.[22]

Our ecological crisis meets these criteria. Abundant uncertainty surrounds the behavior of the climate system. The stakes are astoundingly high. Whether it be global warming, or the destruction of habitats or the erosion of arable land, these are problems of global dimensions, with multiple sources. Tackling them requires taking urgent decisions, with disputed alternatives. We cannot remove the crisis from questions of value involving equity, global justice, justice for future generations and for nonhuman species.

Acknowledging these complexities is a first step toward configuring a response that is as sophisticated as the problem we are trying to address. As biologist Daniel Botkin notes, "[W]e must accept nature for what we are able to observe it to be, not for what we might wish it to be. Accepting this perception of nature, we discover that we have the tools to deal with it."[23]

Can we pull back the veils over our eyes, and force a change of lenses?

CHAPTER 3

ORIGINS

T O UNDERSTAND HOW we find ourselves in such a situation of ecological precarity requires delving into the past. But our intents are fraught from the start. Any attempt to retell human history is held back by the limits of our evidence: the imperfections of written records and the oblivion of oral history. Indeed, much of our archaeological record is so fragile, scattered and sparse that it can be molded to a range of conclusions. Our collective memory is an enigma, vulnerable to our contemporary reading and judgment.

In spite of these ambiguities, we have often strived to hammer out compelling narratives that encapsulate the entirety of human experience. History is summarized by sweeping stories: romantic notions of times when humans were in tune with nature, or dramatic parables of collapsing societies led astray by ecological ignorance. Entire eras are caricatured with broad brushes.

We are also trapped in linear sketches of the past, where history follows a steady mechanical evolution from nomadic tribes to civilizations, from barbarity to enlightenment, from precarity to technology.

As we step across the past, we can begin to look through the prism of complexity, which shows us that history is raucously diverse, dynamic and contradictory, filled with surprises and uneasy conclusions. Given its sheer variety of patterns, mutations and experiences, any simple story is dubious, and any systematic story is incomplete.

The history of how human beings have treated their environment is marked by prudence and pillage, care and callousness, protection and destruction. Societies have demonstrated a range of sustainable and unsustainable practices.[1]

Understanding this history is vital, for the past laid the groundwork for today. Our current climate, charged into change, is the culmination of centuries of environmental abuse. Where do these wounds originate? How can we begin to tell the history of ecological destruction?

Beginnings

Most of our life span as a species has seen us live as hunter-gatherers. On a daily basis, human beings strove for water, food and warmth. Societies saw the abundant gifts of nature, and its extreme perils: floods, famines, frosts, pests, droughts, sandstorms, blizzards, cyclones and lightning. Our ancestors were children of the Earth, immersed in their environments, highly attentive to its rhythms and precarities.

As time passed, societies underwent cyclical transformations. Nomadic groups settled, and then departed. Authoritarian societies shed their form and then regained it.[2] But, over time, many landscapes were increasingly shaped by humans, through hunting and forest cutting. Intensive hunting by humans contributed to the extinction of entire species such as the giant Irish elk, the mammoth and the Gigantopithecus.[3]

As our tools improved, our influence aumented. Fire was one of the first tools that allowed human beings to increase their power in nature, by closing down prey, heating homes and razing the land. Cooked food increased our ability to absorb energy. It is no surprise that many cultures, from the Malagasy to the Madeirans to the Maori, have founding myths associated with fire.[4]

The Neolithic Revolution, initiated over 10,000 years ago, introduced agriculture into human life. Humans acquired newfound abilities to act within nature, selectively breeding plants and using domesticated animals to haul plows and improve the productivity

of lands. Our ecological knowledge grew significantly. We understood that recycling nutrients and applying greater amounts of
water could safeguard soils and boost yields.

Some societies gradually, if inconsistently, moved from nomadic
to sedentary lifestyles. More permanent settlements were erected,
as cultivation became intensive and annual. Populations grew in
areas blessed with life-giving resources, where slow-moving rivers
moved through valleys. These fertile lands, along the Nile, Huang
He and Indus rivers, would host the world's first major civilizations.
This emerging agricultural era would coincide with the beginning
of a remarkably fortuitous era of climatic stability, known as the
Holocene.

In many contexts, farming and settled agriculture had a transformative impact on human societies, given its production of
surpluses of food. Surplus was accumulated wealth—or what we
might today call "capital." In fact, the word "capital" shares the
same linguistic root as "cattle."[5]

The increased storage and hoarding of food surplus proved
pivotal in many contexts. As Ashley Dawson writes: "Much of subsequent human history may be seen as a struggle over the acquisition and distribution of such surplus."[6] Food, the fruit of nature,
became an increasingly privatized good, the source of power concentrated by the wealthy.

Societies were reconfigured as classes of rulers, priests, warriors
and administrators emerged—all linked to the production, protection and control of food surplus. New hierarchies were solidified, as
power and wealth began to flow from the control of nature. The rise
of serious economic disparities coincides with the arrival of agriculture, although not equally across all civilizations, and not in all
societies.[7]

As populations and their required territories grew, more complex political organizations were established to extract revenue and
manage lands. The formation of various states and large political
entities was tied to the rise of taxable agriculture and livestock do-
mestication.[8]

With farming requiring significant human labor, groups of humans were enslaved and put to the land. Wars were waged to conquer territories, and the workforces that lived there. In Sumeria, laborers were put to work to build intricate systems of irrigation that could take advantage of the water flows of the Tigris and Euphrates rivers. Female slaves worked as weavers, as male slaves hauled boats across rivers.[9] Writing began to be used for bookkeeping and accounting. Some of the first written tablets in human history consist of copious records of grain quantities, captives and slaves.[10]

Our early agrarian ancestors began to increase their transformation of landscapes, building infrastructure and expanding slash-and-burn cultivation. In southern Europe, Mediterranean shepherds set fire to forests to clear land for grazing and agriculture.

As states expanded, they required larger portions of land, and more intensive levels of extraction, to sustain themselves. The Sumerian and Egyptian empires began to rely on timber imports. The ample forests of the Levant, North Africa and Mesopotamia were progressively depleted, consumed by construction and naval shipbuilding.[11] Around the time of the Peloponnesian War, Greek cities relied on Sicily and Egypt for half of their food.[12]

The first symptoms of ecological shortsightedness began to appear, as certain communities breached core principles of balance: degrading quicker than regenerating, destroying faster than replenishing.[13]

In ancient Rome, smallholder farming techniques known as *cultura promiscua* were replaced by intensive plantations. Cash crops were grown on major enterprises using slave labor, as maximizing short-term yields became a state priority.[14] Lands across northern Africa were conquered, cleared and fashioned as the empire's granaries. Deforestation spread across the Maghreb, Sicily, Spain and Galilee. The aridity of these regions today is partly a testament to the scale of Roman degradation.[15]

The central environmental problem of ancient times was soil degradation and erosion.[16] In fact, the longevity of civilizations was often determined by their relationship to the earth. Many Sumerian societies ran into trouble as salt accumulated through irrigation and fertility declined. The Egyptian empire, stable over centuries, saw constant fluctuations as a result of erosion triggered by deforestation, deep plowing and irrigation.[17]

In other parts of the world, deforestation gained salience, as industries based on fire—glassworks, saltworks, brickyards, tanneries and smelters—consumed forests.[18]

Amid these challenges, some societies began to call for greater domination of nature. One ancient Sassanid textbook on agriculture spoke of fighting against the desert, conquering it to make it arable.[19] Human beings envisioned even greater projects to sculpt nature and to accommodate it to their needs. Larger constructions, from dykes to dams, were stretched over water bodies. Canals were carved through fields. Hydraulic engineers attempted bolder drainage and irrigation projects, installing drains, sluices and aqueducts. The moors, swamps and plains of Europe were increasingly reclaimed.

In the 14th century, Mongol chancellor Toqto'a attempted to divert the Yellow River using thousands of laborers. In the Chinese region of Nanking, in 1391, over 50 million trees were reportedly planted, to provide lumber for a large sea fleet.[20] In the early 16th century, Leonardo da Vinci, together with Niccolò Machiavelli, tried to divert the Arno River, in an attempt to give Florence access to the sea, and starve Pisa of access. Their plans failed miserably.[21]

The ambitions of power over nature were accompanied by the ambitions of control over greater expanses of territory. Wars took place over lands and water bodies, the sources of life and lucre. As early as the 14th century, Arab historian Ibn Khaldun annotated the pattern of history: conquest, consolidation, expansion, degeneration and conquest.[22]

And, swept up into the extravagance of imperial wealth, various societies practised gratuitous ecological violence. Roman public spectacle, for example, involved unspeakable slaughter. Around

9,000 animals were slaughtered during the inauguration of the Colosseum. Games held to mark Trajan's conquest of Dacia (Romania today) saw 11,000 animals killed.[23] So many lions, elephants, hippos and other creatures were dragged to public arenas and killed that soon they could no longer be found within the empire's bounds.[24]

Notwithstanding these symptoms, until around the 15th century, the mark made by human beings on the biosphere was still minor. The emergence of agriculture, and the communities it gave birth to, still maintained high levels of ecological diversity, self-sufficiency and sustainable productivity. Human communities lived almost exclusively from the energy of the sun.

But two key developments would accelerate the process of ecological destruction: the first was intensified colonialism; and the second was the birth of an economy powered by fossil fuels.

CHAPTER 4

COLONIALISM: THE ACCELERATION

The Conquest was, before anything,
a tremendous butchery.
JOSÉ MARIÁTEGUI[1]

■ ▩ ■ ▩ ■ ▩ ■ ▩ ■ ▩

The invading civilization[s] confused ecology with idolatry.
Communion with nature was a sin worthy of punishment....
Nature was a fierce beast that had to be tamed and punished
so that it could work as a machine, placed at our service
for ever and ever. Nature, which was eternal, owed us slavery.
EDUARDO GALEANO[2]

■ ▩ ■ ▩ ■ ▩ ■ ▩ ■ ▩

It may be stated broadly that the advance of settlement has,
upon the frontier at least, been marked by a line of blood.
LORIMER FISON & ALFRED HOWITT, 1880[3]

■ ▩ ■ ▩ ■ ▩ ■ ▩ ■ ▩

THERE ARE MANY WAYS to see colonialism. A breakneck rush for riches and power. A permanent pillage of life. A project to appropriate nature, to render it profitable and subservient to the needs of industry.

We can see colonialism as imposition, as the silencing of local knowledges and erasure of the other. Colonialism as a triple violence: cultural violence through negation; economic violence through exploitation; and political violence through oppression.[4]

Colonialism was not a monolithic process, but one of diverse expressions, stages and strategies. But its common ethos was a desire to seek access to new territories, resources and laborers. Impelled by God, fortunes or fame, with almost limitless ambition, countries and companies scrambled to acquire control of land. New territories were seen as business enterprises. Local inhabitants were either obstacles to be removed or workforces to be subjugated.

The colonial-imperial era is fundamental to an understanding of how we have arrived here. As Eyal Weizman notes: "the current acceleration of climate change is not only an unintentional consequence of industrialization. The climate has always been a project for colonial powers, which have continually acted to engineer it."[5]

What did colonialism seek? Wealth and power are the abstractions. But concretely it was commodities: metals, crops, minerals and people. Political might, economic growth and industrialization required hinterlands to provide raw materials, food, energy supplies, labor and consumer demand. States sought expansion, appropriating territories and dominions. Between 1400 and 1917, the Russian empire expanded a thousandfold.[6]

Gold and silver supplied the first vice, feverishly obsessing the early colonizers. From the 16th to the early 19th century, around 100 million kilograms of silver were hauled from the mines of Latin America to Europe. The Spanish writer Alonso de Morgado observed at the time that enough treasure had arrived on the shores of Seville by the 1580s to pave the entire city's streets with gold and silver.[7]

Plant commodities—from sugar to spices, cotton to coffee— would follow, as empires arranged the world to satisfy metropolitan tastes. Nature would serve as the canvas, the prize and the victim of colonialist dreams.[8]

The impact on nature

Nature narrates the colonial story, through its vast mines, its desecrated rivers and emaciated territories. Across continents, mangroves, grasslands, rainforests and wetlands were cleared to make way for quarries, plantations, ranches, roads and railways.

As historian Richard Drayton explains, imperialism—the expansion of empire—was "a campaign to extend an ecological regime: a way of living in Nature."[9] Entire landscapes had to be subjected to control and exploitation.

Colonies were arranged to facilitate and maximize extraction. Profit was the compass. French colonial planners divided "useful Africa" from "useless Africa."[10] Lands were surveyed, zoned and parceled. All these endeavors relied on a narrative of emptiness, of nothingness.

The New World's territories were vacant, lands of nobody, *terra nullius*—open for conquest and colonizing. The Arctic, the Outback, the Wild West and the Amazon were (and continue to be) enduring metaphors that allowed colonizers to depict territories as barren wastelands. But these lands were not empty. The fiction of negation, and discovery, was used to justify the clearance of native habitats and inhabitants.

Nature was a blank slate, to be reconfigured and rendered useful. Where colonizers arrived, maps were redrawn, inhabitants ousted, and new methods of production installed. Collective land management practices were shredded, as models of individual property ownership were imposed. New courts and laws governed the territory, handing lands over to concessionary companies and settlers. Long-term residents were now "squatters" on their own land.

Time-tested and locally rooted agricultural traditions were trampled and stamped out.[11] In Mexico, peasants were stripped of their *milpa* lands. In Madagascar, the *tavy* system was outlawed.

Rural areas were dragooned into ambitious imperial strategies, with villages forced to pay tribute or follow new production regimes. Local peasants were subjected to forced cultivation, compelled to grow what they were told. In French Equatorial Africa, the Mandja people were barred from hunting and pushed into work on cotton plantations.[12] In 1905, communities living in the German-controlled Tanganyika (now part of Tanzania) revolted against policy forcing them to grow cotton for export. In response, as historian John Reader recalls: "three columns advanced through the region, pursuing a scorched earth policy—creating famine. People were forced from their homes, villages were burned to the ground; food crops that could not be taken away or given to loyal groups were destroyed."[13] Around 300,000 people would perish.

From continent to continent, staples were replaced by cash crops. Plantation systems were installed, designed to maximize yields. In India, the British entirely reorganized the agricultural system. Lands previously used for low-scale subsistence agriculture, would now be destined for cash crops such as cotton and tea, grown for export to international markets. The Portuguese empire similarly installed cotton regimes across its Brazilian, Angolan and Mozambican colonies.

Communal water management techniques were replaced with enormous works of engineering and state regulation.[14] Pseudo-ecological arguments were often used to discredit local peoples and justify the clearance of communities. Traditional pastoralist practices were framed as outdated, damaging and ineffective. French July Monarchy propagandists used Arab desertification of Algerian land as a justification for conquest: once in control, France would restore ecological order and change the climate.[15]

Perhaps the most destructive agrarian practices involved sugar. In the Canary and Cape Verde islands, sugar production was imposed through deforestation, woodlands were cleared to end up as deserts.[16] The forested Atlantic island of Madeira, which means wood in Portuguese, was virtually stripped of trees to make way for livestock and sugarcane plantations. Slaves, transported from the Canary Islands and Africa, dug thousands of kilometers of canal to irrigate the sugarcane fields. Once Madeira's forests were cleared, and the sugar industry could no longer burn wood to fuel its mills, plantations were replaced with vineyards.[17]

In the Americas, millions of hectares were stripped of forest life and burned to allow for massive cane plantations, accelerating soil erosion. In the West Indies and Guyana, rainforests were demolished to make way for sugarcane cultivation. Haiti, whose name means "green island" in Arawak, was stripped of trees.[18] In Mexico, deforestation exploded with the arrival of the Spanish, as forests were cleared to supply sugar refineries with fuelwood.

The logic of sugar's monoculture was applied to a variety of commodities. The peripheries of the Amazon were cleared for coffee plantations. Using forced labor, Southeast Asia, southern Colombia and the Congo were deforested and converted into rubber plantations. Burma and Thailand saw their forests turned to mass rice-fields, while Indian ecosystems were felled to make way for cotton plantations.

In all these contexts, soils were exhausted and made sterile, degraded by deforestation and monoculture. In areas of Brazil and the Caribbean, the tree-bare terrains left by plantation economies became ideal incubators for mosquitos carrying malaria and yellow fever. Searing epidemics killed major segments of the population.

As historian Corey Ross recalls:

> One of the recurring themes in the history of plantations is the perennial cycle of boom and bust. Whether the crop is sugar, tobacco, or cotton, the basic pattern is often the same: an initial frenzy of clearing and planting is followed by either a precipitous collapse of production or a gradual process of creeping decline before eventually ending in soil exhaustion, abandonment, and relocation elsewhere.[19]

Since there was always more land to conquer and acquire, sustainability was irrelevant. The model was simple: exhaust the land, abandon it and clear new land. But the shortcomings of such short-termist thinking would become readily apparent, particularly in the circumscribed territories in the Caribbean.

In the 19th century, the colonial "discovery" of guano, a fertilizer made from accumulated bird droppings, would temper these concerns around fertility. Guano, the predecessor of contemporary chemical fertilizers, was dozens of times more powerful than domestic manure.

Mania for fertilizers swept across imperial states. Tens of thousands of Chinese laborers were shipped to Peru to mine guano, used to fertilize European soils. A major agrarian transformation was already underway. Previously, agrarian innovation worked to improve crop rotation, enhancing soil fertility by combining livestock raising with agriculture. Yet with the boons of imported fertilizers, agronomists began to envision an agriculture without limits. The balances of rotation were no longer relevant. Nutrient cycling was displaced by nutrient application. As fertilizers gained prominence, and water closets were introduced in elite homes, the recycling of excrement was no longer required.[20]

Fertilizers became the commodities of choice. As guano deposits depleted, saltpeter became the primary source of nitrate. Efforts to control the nitrate-rich Atacama Desert contributed to the War of the Pacific, fought by Chile, Bolivia and Peru between 1879 and 1883.[21] When phosphate supplies were found in the South Pacific, Germany and Britain were quick to annex territories in the South Pacific. Banaba, Nauru and Christmas Island were all strip-mined for phosphates, their local peoples displaced or dispossessed.[22]

In the 19th century, Southern states in the US discovered phosphate deposits. The freeing of slaves coincided with major growth in fertilizer use, as plantation owners replaced their previously press-ganged labor with natural accelerants to cultivate their depleting lands.[23]

▪   ▪   ▪   ▪   ▪

Beyond agriculture, intensive alluvial gold mining in the Caribbean, and silver mining in the Andes and Mexico's Sierra Madre mountains, devastated the local terrain. Trees were ripped out of the ground to fuel smelting furnaces, triggering erosion, flooding and major loss of soil fertility. Around the Bolivian mining city of Potosí, over 30 dams were built to power its mills. But the hydraulic infrastructure installed to amplify production (as well as the local deforestation) caused constant flooding. In 1626, the major San

Ildefonso dam broke; over 4,000 people were killed. Thousands of cubic tonnes of water contaminated by mercury effluent flooded into local rivers. [24]

Loggers also wrought devastating impacts. India's Malabar coast was cleared of teak forests by British merchants. Burma's Tenasserim forest was raided next, stripped of teak over just two decades.[25] Within only a handful of years, Fiji, Hawaii and the Marquesas Islands were cleared of sandalwood. In Canada, settlers set light to forests to provide a core ingredient of potash. In Australia, settlers predicted it would take centuries to clear the "Big Scrub" terrain across New South Wales; it disappeared in just 20 years.[26]

Between 1840 and 1940, Java lost half of its forests.[27] In Madagascar, logging and agricultural expansion cleared 70 percent of the country's primary forests between 1895 and 1925.[28]

From territory to territory, life was swept away. Entire animal species were decimated through overhunting. The demand from European elites for fine furs drove hunters and trappers into Siberia and the Americas, carving open new frontiers. John Astor, founder of the American Fur Company, became the first multimillionaire in US history.[29] Fishing fleets scoured the seas, slaughtering shoals and pods. In less than 30 years, sea cows were harpooned into oblivion across the Bering Strait.[30] Quaggas, thylacines, great auks, passenger pigeons, warrahs and hundreds of other species disappeared within decades. Industrial whaling, driven by demand for blubber, culled whales to the edge of extinction, removing all bowhead whales from the Beaufort Sea.[31]

Those doing the colonizing seemed to be blinded by their lucrative pursuits. The Mexican historian Rafael Bernal pored over the chronicles of Spaniards who came to conquer the Americas. In his studies, he found that remarkably few conquistadores ever described the nature around them in detail. People who had once been farmers from the arid corners of Castilla and Andalucía somehow found little to say about the spectacular sights surrounding them. Nature, like the humanity of those it held, barely existed. [32]

The impact on peoples

Just as environments and animal species needed to make way for productive "civilization," so too did local inhabitants. The eradication and exploitation of nature was conjoined with the eradication and exploitation of peoples. Ecocide came hand in hand with ethnocide. The Guanches, Lucayas, Charrúa and Beothuk are just some of the many peoples massacred on the altar of lucre.

The methods were common: seize, dispossess, exclude, expel, extract and extinguish. Martinican author Aimé Césaire would later note that between "colonizer and colonized there is room only for forced labor, intimidation, pressure, the police, taxation, theft, rape, compulsory crops, contempt, mistrust, arrogance, self-complacency, swinishness, brainless elites, degraded masses."[33]

The life of empire depended on the theft of life. Control of nature, and expulsion of its stewards, often had to be accomplished though gratuitous violence. The friar Bartolomé de las Casas, who had previously participated in massacres, later railed against the greed that "murdered on such a vast scale," executing "anyone and everyone who has shown even the slightest sign of resistance."[34] Wars of clearance were fought to appropriate land, advancing the frontier of settlement and extraction. The British army fought three wars in Burma to gain access to forests.[35] The majority of the population of the Banda islands, around 15,000 people, were either murdered or enslaved. Lush lands were converted into plantations destined for spice production.[36]

Dutch settlers in southern Africa quickly set out to wipe out local San communities, representing a range of Kalahari peoples (the Ju|hoansi, the !Xun, the Heillom, the Nharo and the Khwe). These peoples were portrayed as vermin, destined for either execution or forced labor. Permits were issued allowing settlers to shoot to kill: paramilitary commandos hunted San populations until they were virtually eliminated from the Cape of Good Hope by 1873.[37] Over a century later, David Kruiper, a leader of the ǂKhomani San peoples, would lament: "we have been made into nothing."[38]

In the United States, lands were systematically stolen from

indigenous peoples as the government drove westward, opening areas for settlement. Thousands were expelled from their ancestral homelands, forced down the Trail of Tears. In California, gold rushes, along with their associated massacres and epidemics, decimated the regional indigenous nations. Two decades after the 1849 gold rush, the indigenous population had dropped from 150,000 to 30,000.

Across Australia, similar frontier violence was waged against Aboriginal communities, the "marauding blacks." In the state of Queensland alone, historians Raymond Evans and Robert Ørsted-Jensen have documented the killings of 65,180 Aboriginal Australians between the 1820s and early 1990s. Hundreds of massacres took place. In 1842, on the outskirts of Kilcoy, dozens of members of the Gubbi Gubbi people died after eating flour purposely laced with strychnine. In 1838, at Waterloo Creek, 50 Kamilaroi people were murdered by mounted police, ordered to expel Aboriginal people to make way for farmland.[39] Just four months later, in Slaughterhouse Creek, a further 300 members of the Kamilaroi people were murdered.[40]

Between 1898 and 1902, American troops conquered the Philippines; hundreds of civilians perished. Twenty-six of the American generals in the Philippines had been involved in wars against Native Americans. Brigadier General Jacob H. Smith told his troops: "I want no prisoners. I wish you to kill and burn. The more you kill and burn, the better it will please me."[41]

In Germany's West African colonies, settlers appropriated the lands of Ovaherero and Nama peoples, provoking significant resistance. In response, Ovaherero villages were surrounded, their residents shot, their water wells poisoned. Eighty percent of the Ovaherero peoples, and half of the Nama peoples, were killed. The German general in charge of Southwest Africa (later Namibia) noted: "The natives must give way. Look at America."[42]

In South America, Chile and Argentina handed swathes of Patagonian territory to sheep ranchers and entrepreneurs. At the stroke of a pen, the indigenous peoples who had made the region's plains, inlets and mountains their homes for thousands of years were rendered illegal trespassers. Bounties were placed on them by landholders, rewarding anyone who could deliver the ears, testicles, heads and hearts of indigenous adults and children.

In Chile, the Kawésqar peoples were hunted into oblivion, or deported to missions to die of sickness and sadness. Eleven Kawésqar were dragged away from their homes, to be exhibited at European zoos and public parks.[43]

The isle of Tierra del Fuego was home to thousands of Selk'nam, slowly massacred by mining expeditions searching for gold. One naturalist and military official, Ramon Lista, would direct the massacre of 28 Selk'nam; he was rewarded with a prestigious public position, his name adorning streets. In one infamous case, landowners poisoned a beached whale and hundreds of Selk'nam died after eating the contaminated meat.[44]

In Costa Rica, the Chiriqui Land Company (a subsidiary of United Fruit) expelled the Bribri peoples from the valleys of Sixaola and Talamanca, converting the forest into a valley of banana plantations. In neighboring Panama, a 1904 law talked about the need to "Christianize and civilize the savage Indians," with the result that communities were ravaged and banana plantations extended.[45]

Work and slavery

In the colonial realm, nature and those deemed inferior enough to be part of it had to be put to work. Governed by whips and watches, laborers were forced to work the earth: to slash, mine, break, cut, harvest, extract, carry and cart. Across its centuries, coerced labor found different incarnations, from formal slavery to convict-leasing, from indentured labor to peonage. Under these systems of bondage, human beings were treated as chattel, expendable parts of the exploitation of expendable lands.

Forced labor, and its use to pillage nature, does not originate in 15th-century colonialism. In ancient Greece, hundreds of thousands of slaves forced to work in mines are estimated to have died from lead poisoning.[46] Before Portuguese merchants ever set foot in Africa in the 15th century, up to 10 million Africans had been enslaved and shipped across Europe, Asia and the Middle East over various centuries.[47] In Spain, prisoners and slaves were forced into the pits of Almadén to extract mercury, thousands of metric tonnes of which were taken to the Americas to be used in silver extraction. Slave markets dotted the Iberian Peninsula's cities in the 15th century, selling Muslim prisoners, West Africans, Russians, Greeks, Circassians, Tartars and Guanches.[48]

In the Americas, slavery took hold at the beginning. In 1495, Christopher Columbus's first commercial activity in the Americas involved sending four caravels packed with 550 indigenous slaves, destined for auctions in Europe. Crammed in deprived conditions, 200 of these died on the way and were tossed overboard. [49]

Across the Caribbean islands, colonial cavalry units captured locals. Between 1500 and 1510, thousands of slaves were shipped from the Antilles to the island of Hispaniola to mine gold. Entire islands were emptied and depopulated to fill the gold mines with labor.

Reflecting on his time, Columbus noted: "The Indians of Hispaniola were and are the greatest wealth of the island, because they are the ones who dig, and harvest, and collect the bread and other supplies, and gather the gold from the mines, and do all the work of men and beasts alike."[50] Hernán Cortés, the conqueror of Mexico, owned gold mines and major properties with hundreds of slaves.[51]

Yet slave trading would reach new dimensions over the Atlantic. The extensive plantations of the Caribbean required slave societies, and colonizers turned to the extraction of black labor. Human trafficking forced around 12 million human beings, mainly from western Africa, into crammed ships crossing the Atlantic. When the slave trade was at its peak, West Africa's population fell by 20 percent between 1700 and 1820.[52]

Manacled, shackled and chained, Africans were transported in fetid floating prisons across the seas. Olaudah Equiano, a survivor of the journey, recalled that "[t]he shrieks of the women, and the groans of the dying, rendered the whole a scene of horror almost inconceivable." At least a third of slaves died en route.[53] Those who did not perish during the Middle Passage were sold at auctions in the Americas, where bodies of slaves were standardized and commodified. Devastating distinctions were drawn as humans became merchandise. The value and identity of human beings were defined by their financial value: "Extra Men, No 1 Men, Second Rate or Ordinary Men; Extra Girls, No 1 Girls, Second Rate or Ordinary Girls."[54]

In Mexico, indigenous and African slaves worked side by side in the silver mines of Parral.[55] In the Andes, subjugated indigenous populations were handed forced labor quotas by the conquerors. Over centuries, millions of forced laborers were drafted, dragged from their communities and sent into the living hells of mines such as Potosí, where toxic mercury used to process ore poisoned both miners and their communities.[56] Such systems of forced labor were only formally abolished in the 1820s.

▪ ▪ ▪ ▪ ▪

Millions of slave laborers were not enough, and eventually the pressures for slave emancipation would push commercial interests to source cheap labor under different arrangements. In the United States, the 13th Amendment abolished slavery for all citizens, with one exception: criminals. In the wake of abolition, penal systems reconfigured their subjugation. Convict-leasing secured a continued source of cheap labor for plantations.

Boatloads of indentured laborers were brought over from China to mine guano and gold in Peru. Between 1846 and 1932, around 24 million Indians were shipped out of the subcontinent as indentured laborers, heading to the Caribbean, East Africa, Mauritius, the Guyanas, Melanesia and Micronesia.[57]

As in many states across the world, the prison system was tied to colonization and continued exploitation. The British East India Company sent hundreds of Indian convicts to colonize the Andaman Islands, Mauritius and Southeast Asia.[58] The British government sent prison hulks, ships filled with prisoners, to supply the labor needs of Bermudan shipyards. Perhaps its most ambitious use of prison labor was its establishment of penal colonies in Australia, but this time it filled them solely with convicts of European origin: 165,000 men and women were transported. Once in Australia, they served on chain gangs building public works.

With rising demand for bicycles and the arrival of the automobile, rubber became the new boom commodity. As demand soared, Brazil became the hub of production, with the Amazonian capital Manaus growing into one of the wealthiest cities in the world.[59] In the Belgian Congo, brutal delivery quotas for rubber meant that thousands were worked to death; if villages failed to meet their production quotas, it resulted in flogging, torture or execution.[60] In Colombia's Putumayo Valley, frenetic rubber traders maimed dozens of indigenous communities. The rubber baron Julio Cesar Arana, founder of the Peruvian Amazon Company, used debt peonage to enslave thousands of indigenous laborers. Estimates suggest that tens of thousands of people were killed under his administration.

The violent suppression of any dissent was endemic across plantations and extractive industries. In 1907, thousands of nitrate workers were mowed down by the Chilean army in the city of Iquique after striking to demand better working and salary conditions. In April 1912, goldfield laborers in northeast Siberia working near the Lena River went on strike over extreme working conditions, only for hundreds to be killed by the Imperial Russian Army. In 1942, the Bolivian army opened fire on the camps of striking tin miners in the village of Catavi, killing dozens. In 1928, 80 banana plantation workers in the town of Ciénaga were killed by the Colombian army, after a union strike demanding better conditions.

But despite the stakes, as colonial states waged violence, and extractive companies ventured into territories, local inhabitants responded. From mutinies on slave ships to insurrections on plantations to anticolonial revolts, rebellion was the rule. In 1595, runaway slaves in São Tomé and Príncipe took control of the capital, razing sugar mills across the territory. From Ghana to Kenya, new land bills met with fierce local resistance. Across the Americas, Maroons established free communities, resisting colonial incursion.

Slavery was violently resisted, from the Pueblo Revolt of 1680, to the major slave rebellion across the French colony of Saint-Domingue in the 1790s, which liberated the territory that became Haiti.[61]

The destruction of memory

The consequences of colonialism and imperialism, in all their forms and across all their epochs, defy our imagination. Unspeakable cruelties were inflicted, bearing unspeakable scars and agonies.

Colonialism was, and remains, a wholesale destruction of memory. Lands, the sources of identity, stolen. Languages, ripped from mouths. The collective loss to humanity was incalculable, as cultures, ideas, species, habitats, traditions, cosmologies, possibilities, patterns of life and ways of understanding the world were destroyed. Countless ecological traditions—involving diverse ways of being with nature—were swept away.

As formal colonialism came to an end, the process of erasing its crimes from public memory and effacing history began. The forces of forgetting crafted and promulgated mythological narratives of innocent imperial greatness, unblemished by enslavement or genocide. When forced to give away the Congo, King Leopold took to burning all documents associated with his brutal rule. "I will give them my Congo, but they have no right to know what I did there," Leopold said. His palace's furnaces burned for eight days.[62]

There are many such shredded chapters that we will never reconstruct. Every death count, every statistic, every fragment of the past, is bitterly incomplete. But the preliminary arithmetic of cruelty is enough to illustrate the sheer magnitude of destruction.

So catastrophic and widespread was the decimation of human life in the Americas that nine-tenths of its original population was extinguished through war, epidemic diseases, enslavement, overwork and famine.[63] Most of us have heard the simplistic story of a genocide by germs, where populations were wiped out by diseases to which they had no immunity. But the vulnerability of communities to maladies was not just a product of biological misfortune.

Malnutrition, exhaustion, absent sanitation, enslaving missions and overcrowding helped to weaken people's protection.[64] Demographic research has shown, for example, that on Hispaniola the indigenous population plummeted before any smallpox cases were documented.[65]

In the last decades of the 19th century, tens of millions of Indians died of famine, while British colonial policy forced the country to export record levels of food. If their bodies were laid head to foot, the corpses would cover the length of England 85 times over.[66] The evisceration of the Congo, designed to extract maximum levels of ivory and rubber, killed at least 10 million people—half the country's population at the time.[67]

The bounties of colonialism underwrote the wealth of Europe. Seams of silver and gold swelled the coffers of banks and merchants. The fortunes made from metals, slave trading and plantation commodities served as direct stimuli to colonial economies, helping to bankroll the Industrial Revolution.[68] Consumers in the colonies proved vital to purchasing products and fueling the rise of western European industries.[69] By the late 19th century, over half of the British state's revenue stemmed from its colonies.

Colonialism reconfigured the world economy. India's share of the global economy shrank from 27 percent to 3 percent. China's share shrank from 35 percent to 7 percent. Europe's share exploded from 20 percent to 60 percent.[70] The tables of development were overturned. In the 18th century, differences in income across the world's leading civilizations were minimal. It is in fact likely that average living standards in Europe at this time were lower than elsewhere.[71]

At this time, manufacturing districts along the Yangtze delta and the Bay of Bengal rivalled the artisan workshops of Florence. Historian Prasannan Parthasarathi argues that "there is compelling evidence that South Indian laborers had higher earnings than their British counterparts in the 18th century and lived lives of greater financial security." At the start of the 19th century, Europeans survived on calorie levels below the average in Latin America and North Africa.[72] But through pillage and conquest, the game changed.

The story of colonialism, sanitized and blotted out from the historical consciousness, needs to be recalled, for many reasons—not the least of them because of our concerns about the climate. Colonialism's ledger of lavish destruction—its wholesale removal of ecosystems, and the subjugation of those communities that had nourished them—unleashed major rises in emissions. Between 1835 and 1885, deforestation in the territories of the United States was the largest global contributor to emissions.[73]

Ultimately, colonialism transformed the speed, scope and scale of ecological destruction. It took two centuries for 12,000 hectares of forest to be felled in medieval Picardy while it took only a year for an equivalent amount to be cleared during the Brazilian sugar boom of the 1650s.[74] Dramatic changes in land and marine ecosystems were normalized, and the dynamics of economic growth were altered with significant implications. Political ecologist Jason Moore argues that "the rise of capitalist civilization after 1450, with its audacious strategies of global conquest, endless commodification, and relentless rationalization," marked "a turning point in the history of humanity's relation with the rest of nature, greater than any watershed since the rise of agriculture and the first cities."[75]

Across most continents and contexts, the grip and influence of empire impelled an era of major devastation. As environmental historian Joachim Radkau outlines, "[i]n the opinion of the vast majority of scholars, a large-scale ecological crisis developed in the 18th century and became acute and obvious in the 19th.... In China, as in Europe, one can detect in the 18th century a desire to use natural resources to their limits and to leave no more empty spaces."[76]

Its legacies endure today in colonial complexes that underlie our visions of nature, knowledge and other humans. Economically, its inheritance was the naturalization of a model of intense cost-shifting, which allowed for states to offload resource-consuming industries, and the costs of ecological damage. By the birth of the

New World, silver mines and seams in Bohemia and Saxon had been exhausted. European forests were bearing the burden of centuries of exploitation for use in shipbuilding. Around 3,000 oaks were required to build a single warship.[77] Iberian shipbuilding, which had eaten through the forests of Catalonia, was transplanted to Cuba and Brazil.[78] The construction of British battleships was transferred from London to Bombay shipyards.[79] Once the industries had been externalized, resources could be extracted with scant attention paid to the environmental consequences. Japanese policies, for example, protected forests in Japan, but exploited them during Japan's rule of Korea.[80]

Colonialism also firmly shaped the ways we view conservation and ecology. Colonial efforts to protect nature, particularly popular at the end of the 19th century, became further opportunities for colonial control. Inhabitants were removed from areas of "pristine nature" that then became national parks, while lands outside these were devoted to intensive extraction. Ahwahneechee communities were, for example, expelled from the valleys that today make up Yosemite Park in California.

Colonialism within countries

Empire—its expansion, and its trampling of life—was a project that pervaded ideologies. While western European empires paved a particular path of extreme ecological destruction in their colonies, they were not its sole practitioners. In India, coercive state control of forests predated the arrival of the colonialists. The Mauryan state and Mughal empire conducted clearing and settlement. Thousands of people were deported from Kalinga, and sent to establish new settlements.[81] Irrespective of their origin or political grounding, many civilizations from the Asante Kingdom to Qing China exhibited ecologically destructive traits.

Neither was ecological evisceration reserved for distant colonies. Within Europe, major energies were expended to remake "rural areas" and subjugate peasant peoples to the productivities required for participating in an age of increasing trade.[82]

Scandinavia, Poland and the Baltic states were turned into saw-mills: sources of timber for the wood-fuelled empires of the west of the continent. The rates of deforestation in Poland in the 16th and 17th centuries were unparalleled, comparable only to what was occurring in northeast Brazil. Poland and the Baltic states also became granaries for Europe. So strong was the agrarian shift that, in the 17th century, the number of people living in cities in Poland fell.[83]

Across Europe, the velocity of the transformation of nature shifted dramatically. European metal production quintupled between the 1450s and 1530s.[84] Across Eastern Europe and Central Asia, entire forests were cleared. The conquered steppes of southern Russia were rapidly overexploited, their lands despoiled through erosion. Entire riverbanks were stripped of trees, laying the ground for mass flooding. In 1879 the Maros River overflowed its banks and swept away the Hungarian city of Szeged.

The era of nation-states also involved extending control over peoples and natures. With differing degrees and intensities, peasantries were converted, absorbed and assimilated into state structures. National identities were forged and inculcated. Dominant languages were standardized, alphabetized and imposed. National territories were woven together by road and rail.

Across Scandinavia, Sámi populations were subjected to colonization, state control and missionary activity. Sámi village districts (siida) were replaced with settlements, their lands appropriated by state administrations.[85]

A defining feature of this era was the enclosure of land, which gained new force in Renaissance Europe. Enclosure acts, game laws and legal instruments were introduced to brutally separate rural populations from the means of subsistence and survival. Rights to access the "commons" were shredded. Anti-poaching laws prohibited hunting. Peasants, previously self-provisioning, were turned into laborers.

Nobles shredded the right of habitation, seizing and privatizing previously common lands. Those flaunting the new legislation were sent to prison: in the UK, for instance, a fourth of those condemned to prison between 1820 and 1827 had committed the crime of poaching.[86] Decade by decade, new legal instruments swept lands into the hands of elites. Between 1760 and 1870, parliamentary acts enclosed one-sixth of England. Open fields were fenced and hedged to allow more intensive agriculture and private enterprise.[87]

Historian Joan Thirsk noted that "after enclosure, when everyman could fence his own piece of territory and warn his neighbors off, the discipline of sharing things with one's neighbor was relaxed, and every household became an island to itself."[88]

Peasants who had previously been self-reliant in their rural environments were cajoled into urban wage labor in order to buy commodities necessary for survival. Most despised this coerced precarity. Draconian vagrancy laws, aimed at forcing peasants into submission, were introduced from the Low Countries to Switzerland.[89] By the year 1700, landlords had acquired two-thirds of farming land.[90]

The drastic changes generated fierce opposition. Peasant revolt exploded across Europe. Priest Thomas Münzer denounced the process as one where "every creature should be transformed into property—the fishes in the water, the birds of the air, the plants of the earth: the creatures, too, must become free." Travelling around 18th-century France, agriculturalist Arthur Young remarked, "I know of no country where the people are not against enclosures."[91] But the revolts were widely repressed and their leaders executed.[92]

Over centuries, in different forms, and at different paces, enclosure swept across the world, installing new regimes of nature. In Scotland, the Highland Clearances saw small-scale farmers removed from their land, and replaced by privatized, ranch-style farming. Gaelic speakers and Highlanders, judged inferior and backward, were displaced. In 17th-century Ireland, English authorities tore up the customary *Brehon* laws, which enshrined landholdings in kin-groups (*derbfine*), instating entirely new land regimes.[93]

Across Russia, the *mir* system of communal land management was stripped away by tsarist policy in the early 20th century. In 1904, the historian Vasili Kliuchevsky described Russia's story as "the history of a country that colonizes itself."[94] The first modern genocide in Europe involved the massacre of the Circassians by the imperial Russian army, and the destruction of their *auls*, small villages ruled by communal councils.[95] The lands of the Karachais, Balkars, Nogais, Kalmuiks, Vespians and Shors were similarly assimilated. In Russia's Yamal Peninsula, Nenets, Khanty and Selkup peoples made up over four-fifths of the population in 1897. Subjected to persecution, collectivization and expulsion, within 40 years they had become minorities. By 1989, they represented less than eight percent of the population.[96]

At the start of the 20th century, Japan's rulers sent thousands of convicts to the island of Hokkaido to clear land, build infrastructure and displace the indigenous Ainu populations.[97] Many Ainu were enslaved; their language prohibited.

Even states defying colonial rule adopted such approaches. In Ethiopia, the central government strengthened its grip. As environmentalist Tewolde Egziabher explains: "The preoccupation of the Ethiopian government was one of maintaining central power so as to withstand colonial attempts from Europe. To this end, the central government has seen local autonomous movements as dangerous and therefore traditional systems of land management...were systematically destroyed."[98]

A Cold War consensus

The relentless impulses toward control would gain even greater momentum when fuelled by the technologies and intellectual currents of modernity in the 20th century. The latter half of this period is often simplistically depicted as the struggle between communism and capitalism. But these two conflicting ideologies shared a common disregard for human reliance on ecosystems. Soviet ideology, for example, was underpinned by visions of dominating nature. Leon Trotsky believed that "man...will learn how to move rivers and mountains, how to build people's palaces on the peaks of Mont Blanc and at the bottom of the Atlantic."[99]

Behind the grip on state power was a stranglehold on nature. Countless projects to impel development and control nature were implemented. Stalin's first five-year plan included a doubling of coal production.[100] Peasant traditions were eviscerated, as farm collectivization imposed state control on landscapes.

Rural development policies were guided by the ideas of agriculturalist Trofim Lysenko, who condemned "bourgeois biology" for merely aiming to understand nature when the goal was to radically change it.[101] Lysenko heralded the powers of science to transform reality: plants could be educated.

Any dissent to the new biological ideology was crushed. Hundreds of agricultural scientists were imprisoned. Many were either executed or starved to death.[102] In 1940, as he collected plant samples in Ukraine, the prominent geneticist Nikolai Vavilov was arrested. Interrogated for months on end, he died of starvation in prison while his research colleagues were killed.[103]

Once again, the law of environmental destruction took hold: those deemed disposable were enslaved and put to work taming nature. Toward the late 1920s, Soviet security organs began to shoulder responsibility for economic development. Secret Police chief Genrikh Yagoda crafted a new penal system composed of "colonization villages." These villages would be filled with detainees and positioned in remote regions of the country, strategically chosen to expand farmland and maximize the extraction of key resources. Prisoners would help, in Yagoda's words, "to colonize the north at the fastest possible tempos."[104] Gulag prisoners were effectively the pioneers forced to expand the USSR's extractive frontiers, and to erect its most dangerous public works.[105] Millions of prisoners were deported and put to work carving canals, mining coal, laying railroad tracks and damming rivers. Labor brigades were sent into remote areas to extract raw materials. This mass of bodies, nicknamed "white coal" by their guards, made up in part for the technology the USSR lacked.[106] As economic historian Paul Gregory observed, "Stalin presumed that surpluses could be extracted through Gulag labor." Gulag prisoners accounted for between 85 to 100 percent of the laborers in Soviet gold, diamond and platinum mines.[107]

Between 1929 and 1953, around 18 million people passed through the Gulag system.[108] Emaciated and sickened, they were human oxen, yoked to the state's motor of growth. One Gulag division, the Dalstroi, was tasked with extraction across a territory four times the size of France.[109] Gulag laborers proved instrumental for both the Soviet nuclear programme and the opening of major coal extraction frontiers.[110] By the 1970s, the Soviet Union led global coal production.

Mass tragedies were routine. Tens of thousands of laborers died constructing the Belomor Canal. In one case, 6,000 peasants were transported to the Siberian island of Nazino; within three months, 4,000 of them were dead.[111]

Restrictive agricultural regimes were installed across Ukraine and Kazakhstan, precipitating "extermination by famine," with millions of fatalities.[112]

In the late 1940s, Stalin promoted his Great Plan for the Transformation of Nature. Nature was to be engineered and sculpted into the machinery of economic growth. Rivers were to be straightened and dredged, old forests clear-cut while new forest belts would be planted to prevent drought. The result was devastating. Rivers shrank, fish populations were decimated, and swathes of land turned into "industrial deserts." Cellulose production polluted Lake Baikal.

Siberian author Valentin Rasputin lamented:

> Wherever dams are put up and reservoirs swell, a river ceases to be a river and becomes a disfigured beast of burden with all the life squeezed out of it. After that, the river contains no fish, no water, no beauty. Energy will begin to draw industry,

industry will demand new energy, then more industry will move in, and so on until the Katun, its banks, and the land far around will simply vanish into thin air.[113]

Perhaps the most emblematic case of ecological folly took place in Uzbekistan, where the Amudarya and Syrdarya rivers were diverted to help irrigate massive cotton fields. Millions of liters of insecticides, pesticides and herbicides were sprayed to maintain the cotton monoculture. Within decades, the Aral Sea—the fourth-largest inland sea in the world—had dried up and largely disappeared. The Karalpak, a seafaring people who had inhabited the area for centuries, saw the home of their culture and livelihoods dissolve before their eyes. The copious use of agrochemicals, and the sea salts exposed through desiccation, contributed to immense pollution and dust storms in the area. The region around the Aral Sea soon became the region with the highest infant and maternal mortality in the Soviet Union.[114]

From Stalin's economic plans to Khruschev's Virgin Lands scheme, there was a procession of grandiose projects geared to taming nature. By the time of the Soviet Union's collapse, half of Russia's surface water was polluted.[115]

Territories under Soviet dominion were equally embroiled in ecological neglect. In Poland, environmental information was classified as a state secret. The GDR emitted the highest levels of carbon dioxide per capita in the world by the late 1980s.[116]

In Maoist China, meanwhile, the leadership saw conquest of nature as essential to the solidity of communism. "Man must conquer Nature," a slogan proclaimed. Transferring resources and relocating populations, China's leadership envisioned a drastic transformation of the national economy.

The Great Leap Forward unleashed deforestation and upheaval across China's countryside.[117] During the Great Pests Campaign, citizens were called upon to exterminate rats, flies, mosquitoes and sparrows. This carnage neglected the vital role played by these species in maintaining the ecological balance.

The state ripped up traditional models of agrarian management, taking centralized control of food production and allocation. Grain allotments were disproportionately destined to cities. In one of history's largest single incidents of mass murder, around 36 million people died from famine between 1958 and 1962. In rural villages, famished villagers stripped trees of all their leaves and bark. Terror was readily deployed to acquire food; campaigns against hoarding resulted in beatings and executions.[118]

Singular agrarian models, based on "extolled" communities such as the Dazhai agrarian commune, were promulgated as universal templates. Terracing was launched across the country, irrespective of geographic conditions. Forests and pastures were destroyed to make way for terraces. In some cases, artificial hills were even built in plains, merely for the sake of terrace construction.[119] To supply agrarian projects, major irrigation infrastructure was established; between 1957 and 1959, water extraction from the Yellow River increased by 83 times.

Scientists critical of major development projects were silenced and purged. The hydrologist Huang Wanli, for example, a critic of mega-damming projects, was sent into forced labor.[120]

Water engineering, intrinsic to Chinese civilization for millennia, was taken to a new level. Tens of thousands of dams were built, often with little oversight and planning. Millions were displaced. In 1973 alone, over 500 dams collapsed. In 1975, Typhoon Nina barreled across Henan province, overwhelming the Banqio Dam. Around a quarter of a million people died in the wake of the onrushing waters.

Mao's schemes left a wake of environmental destruction: desertification, forest loss, erosion and a continued commitment to intrusive projects. In the post-Mao period, the highly controversial Three Gorges Dam was rushed through amid the political convulsions following the Tiananmen massacre of 1989; its construction displaced one million people. The intensive development models of the Chinese state have caused over 28,000 rivers to disappear over the past two decades.[121]

This obsession with the conquest of nature was, however, as we have already seen, by no means exclusive to communist states. With new technologies, human beings could embark on increasingly ambitious projects to shape nature. Forced laborers spent 10 years excavating the Suez Canal, for instance, with thousands perishing in the process. Entire valleys of earth were removed to make way for the Panama Canal that united the Atlantic and Pacific Oceans; in the process, over 30,000 laborers died.[122]

The great forces of water were harnessed by engineers to power development. Major dams were stretched across the Niger, Volta and Zambezi rivers in Africa. Local visions of water use and management were pushed aside. From Nehru to Nasser to Roosevelt, major hydroelectric dams were the prestige projects of choice in the 20th century. The downstream consequences were secondary.

In Egypt, tens of thousands of Nubian families were displaced by the Aswan Dam. The dam's agrarian impact was immense. The stability of Egyptian civilization rested on the fertility of its riverbanks, on the annual flooding of the Nile that would deposit nutrient-rich silt on its banks. But the Aswan Dam put an end to the cycle—salinization increased, farmers began to depend increasingly on agricultural chemicals, and the fertility of the land plummeted.

Violence and technology

During the 20th century, tight relationships began to be formed between violence and the quest for control over nature. As imperial states vied to extend their powers, science was brought into the military, applied to accelerate and improve the machinery of violence.[123]

War drove technological change, and technologies forged during war efforts were quickly applied to the destruction of nature in peacetime. Gunpowder and dynamite were brought into mines. Nylon, developed to displace Japanese silk in the stitching of military fabric, was used to stitch massive fishing nets kilometers long. Fishing ships used military sonar, radar and acoustic detection to track fishing shoals. The trampling power of tanks was extended to bulldozers. Chlorinated gases developed to choke soldiers could now be used to choke pests. Many of the drivers of our ecological crisis (pollution and greenhouse-gas emissions) saw almost exponential rates of growth after World War Two.[124]

War drove the appetite for velocity, and the energies offered by fossil fuels. When World War One began, the British Expeditionary Force in France had a mere 827 motorcars. By the end of the war, the British army had 56,000 trucks, 23,000 motorcars and 34,000 motorcycles.[125]

Nuclear bombs, the apex of human capacity for destruction, were widely legitimized as instruments for exploiting nature. In 1949, Yakov Malik, the Soviet ambassador to the United Nations, justified nuclear tests by claiming: "We want to put atomic energy to blowing up mountains, changing the course of rivers, irrigating deserts, laying new lines of life there where the human foot has rarely stepped." From France to the United States, governments conceived of the bomb as an instrument to alter and control nature in radical new ways. Proposals to build canals across Colombia and Panama using hundreds of nuclear bombs were reasonably discussed. Experiments were conducted across North America, using nuclear explosions to extract oil and gas.[126]

The German chemist Fritz Haber, one of the architects of modern industrialized agriculture, exemplified this relationship between control over nature and violence. Crops need nitrogen to thrive. In the 18th century, as we have already seen, nitrogen inputs for exhausted lands began to be supplied through guano and nitrates imported from South America. At the start of the 20th century, the depletion of fertilizer reserves raised fears around food production. In 1909, Haber and engineer Carl Bosch pioneered a method of converting nitrogen into ammonia so it could be used as fertilizer. A new industrial production process began, generating huge amounts of fertilizer. Soon after, Haber began experimenting with poison gases, for both military and economic purposes. His research into chlorine gas led to its use in the trenches of World War One. In the wake of the war, Haber was adamant about his desire to "turn the means of extermination into sources of new prosperity."[127]

Research expanded into pesticide gases. A wide network of research institutes and centers arose, focused on the practical applications of toxic gases. They developed cyanides for pest control that could be used without harming those dispersing it. Cyclon A was the first substance. In the 1940s, Zyklon B, derived from a pesticide pioneered by Haber's institute, was used to kill millions in the gas chambers of the Nazi death camps.[128]

Neocolonialism: The metabolism of misery

During the 19th and 20th centuries, formal colonialism came to an end. Countries were liberated, new flags unfurled, and rewritten constitutions adopted. But although imperial states had been forced to relinquish their hold, their legacies prevailed. Centuries of enslavement, despotism, crushed sovereignty and ecological demolition had guaranteed a long afterlife to imperial haunting, and its logics of conquest and predation. Many of the new nation states carried on down tracks laid for them by the colonial powers and continued the process of environmental destruction. Under the banners of development, thousands of communities were evicted and displaced in development programmes.

In India, between 1947 and 2000, around 24 million Adivasis (indigenous peoples) were displaced by large development projects. The construction of the Narmada Dam displaced over 100,000 people alone. In Brazil, military and nonmilitary governments triggered the wholesale destruction of huge areas of the Amazon rainforest, subsidizing road building, clearing the way for large cattle ranches and opening up the land for migrants. In Egypt, the regime of Hosni Mubarak transferred control of land to large landowners, evicting hundreds of thousands of farmers under the banner of "development."

In 1972, following colonial precedents, the Nigerian government outlawed traditional agriculture by fire clearance, a move that would subsequently contribute to devastating famines.[129] In addition, the government's encouragement of new oil projects was described by prominent Ogoni leader Ken Saro-Wiwa as "recolonization."[130]

Deforestation took hold across former colonies. Between 1960 and 1980, Indonesia's timber exports rose 200-fold. Côte d'Ivoire's timber exports rose from 42,000 tonnes in 1913 to 1.6 million tonnes in the early 1980s; less than a fortieth of the country's forests remain.[131] Between 1900 and the present day, over half the "developing world's" forests were removed.[132]

Those resisting these models were met with severe repression and extrajudicial violence.[133] This metabolism of misery continues to this day, with hundreds of social leaders and community activists killed worldwide every year, for resisting the encroachment of extractive endeavors. Between 2010 and 2016, at least 124 environmental and land activists were murdered in Honduras.[134]

The frontiers of ecological destruction are constantly expanding, as the global economy's appetite for new materials staggers on. Between 2003 and 2015, the number of mining projects in Argentina rose from 40 in 2003 to 800 in 2015.[135] A fifth of Peru has been conceded to mining companies.[136]

Today's world is a landscape scarred by environmental violence: the monocultural soybean fields of Brazil's Mato Grosso; the modern gold rushes of Madre de Dios and Zamfara; the vast tar-sands ponds of Canada; the forest-consuming coal mines of Kalimantan; the megadams of the Mekong Delta; the rivers dredged to yield sand; the phosphate mines of Western Sahara; the palm plantations of Tela; the bauxite mines of Guinea; the mesh of pipelines across the Niger Delta; the sugarcane fields of Uttar Pradesh.

It is also a world of furnaces: the brick kilns of Peshawar; the smelters of Norilsk; the glass industries of Firozabad; the chemical factories of Dzerzhinsk; the steel mills of Xingtai and Mandi Gobindgarh; the fertilizer plants of Baocun; the tanneries of Hazaribagh and Rawalpindi; the aluminium smelters of Al Jubail; the polluted deltas of Ogoniland; the ship graveyards of Bangladesh; the cancer villages of industrial China.

The full impact of colonialism would be revealed in its long-term impacts. It radically transformed landscapes, state relations, philosophies and cultures, leaving as one of its principal inheritances an intensive and plunderous economic model. In pursuit of resources, countries ran roughshod over limits, and destroyed many of the ecosystems and human interactions necessary for preventing climate change.

CHAPTER 5

FOSSIL FUELS,
FURIOUS FLAMES

L IGHT MAKES LIFE. Our nearest star, the Sun, is our central power station. Currents and winds distribute that energy across the world. Energy sweeps through ecosystems, stabilizing temperatures, feeding plants and, in doing so, feeding all the relations of life across the planet. Solar heat drives the water cycle, evaporating water into rain, which nourishes plants. Through photosynthesis, plants build with light. They take carbon from carbon dioxide, add electrons from water molecules and emit oxygen as waste.

Fossil fuels—oil, gas and coal—are deposits of extinguished life. Over hundreds of millions of years, the remains of ancient plants and animals decomposed, and were fossilized in the Earth's crust. Fossil fuels are effectively compressed sunlight, the burial of millions of years' worth of photosynthesis underground.

Fossil fuels have been used by human beings for thousands of years. Around 40,000 years ago, Neanderthals in what is now Syria used bitumen to polish stone tools. Ancient Egyptians used oil for medicinal purposes. In Venezuela, indigenous peoples used crude oil and asphalt for illumination and to strengthen canoes. In Sumatra, "earth oil" emanating from seepages was used to relieve stiffness in the limbs.[1] Across the Middle East, oil was used in medicine, road construction, architecture and shipbuilding.[2] In Persia, oil seeping from the ground was used to bind bricks and caulk boats.[3] Small-scale oil and gas gathering took place across Chinese and Burmese villages; the Chinese city of Kaifeng, for a brief period around the 11th century, used coal for heating purposes.[4]

But overall, extraction of such fuels was minor. Many communities saw sanctity in the combustive power of hydrocarbons. In Azerbaijan's Absheron Peninsula, burning gas leaks from pores in the ground, and the peninsula is a holy site for Zoroastrians, who have flocked for thousands of years to worship the rocks and the holy flames they emanate.[5]

In Central Java, the Manggarmas flame has been the site of sacred Buddhist ceremonies for centuries. At Chimaera in Turkey, the site of a burning gas seep, ancient Greeks built the temple of Hephaestus, god of fire. In Colombia's snow-peaked highlands, U'wa communities regarded oil as *ruiria*, the blood of the Earth.

Only beginning in the 18th and 19th centuries did fossil fuels gain prominence, as they began to be located, extracted and combusted at scale.[6] The carbon cycle, the balance of absorption and release, was radically disrupted. Fossil fuels are deep sources of carbon dioxide, and their increased burning issued huge bursts of energy into the Earth system. Today, we continue to combust fossil fuels at alarming rates, pouring carbon dioxide into the atmosphere at a rate 14,000 times faster than natural cycles over the past 600,000 years.[7]

The fossil economy did not arise irresistibly, out of nowhere. It had to be imposed on others. In India's Khasi Hills, for example, local communities had historically made amulets from coal. Although aware of coal's combustive potential, they were indifferent to its large-scale extraction. Colonial administrators were shocked. In 1810, entrepreneur William Jones opened up the Raniganj coalfield. Local villagers were forced into the pits under threat of expulsion. Within three decades, Raniganj's production had swelled by 20 times. In Borneo, meanwhile, military squadrons took hold of the coalfields of Labuan, in which the local population had been similarly uninterested. The first governor of Labuan reflected: "to do any good, the natives must be controlled." Within decades, thousands of tonnes were being shipped annually from the Malay archipelago.[8]

Gradually, colonial ruling classes took steam power across the globe. As historian Andreas Malm describes:

> Fossil fuels are by their very definition a condensation of unequal social relations, for no humans have yet engaged in systematic extraction of them to satisfy subsistence needs... [S]team power was explicitly conceived as a weapon to augment the power over the peripheries, haul in the products of all continents, dispatch manufactured goods in return, and ensure military superiority all along the way...the British Empire went about subjugating the world to the logic of the fossil economy—a novel structure, utterly absent from all but a tiny corner of the canvas of history, making its way over the surface of the earth during the 19th century.[9]

From Natal to Nanking, empire and enterprise scanned and scoured the planet for fossil fuels. Laborers, forced and indentured, were sent into the depths of the Earth. Slaves and, later, leased convicts dug up coal across the Appalachian mountain range.[10]

In Nigeria, the British established coal mines in Enugu and the Udi Hills, to source fuel for machinery, transport, waterworks and electricity generation. In 1914, various towns in Enugu rebelled against corruption, forced labor and working conditions on road and railway projects. In response, the British colonial authorities opened fire on protesters.[11]

And as workers realized the strategic importance of coal to the economy, they organized. Colorado and West Virginia, centers of coal mining, saw constant face-offs between labor and ownership, ending in the Ludlow Massacre, when armed forces killed two dozen striking coal miners at a tent city.

The exceptionality of fossil fuels

Across all these coalfields and countries, new economic dynamics were emerging. Economic wealth and power began to be drawn from the burning of fossil fuels. The foundations of a fossil economy, powered by fossil energy, were put in place. This economy used the energy surpluses of stored sunlight to accelerate its growth.

The energies we have relied on have profoundly shaped our societies. In the early stages of human history, we relied on the power of human muscles, on the heat of burning wood, on the currents of rivers, on the heft of winds. Over time, we increased our ability to capture energy. Human invention crafted wheels, pulleys and sails: technologies that enhanced the energy we could retain and utilize.

One way of understanding energy is to express it in terms of watts. Energy historian Vaclav Smil compares the profiles of diverse energy types: an average human being can perform around 60–100 watts of work in a day; domesticated livestock 250–800 watts a day; sails and waterwheels 2,000–4,000 watts; windmills 1,000–10,000 watts. The steam engine (100,000 watts a day in 1800) then gave way to the steam turbine (25,000,000 watts a day in 1914).[12]

Where this greater intensity of energy was sought, environmental pressures mounted. The intensive use of wood brought deforestation. The use of whale fat for illumination decimated Atlantic whale populations. Starting in the late 18th and early 19th centuries, fossil fuels became the engine of the global economy. Combusted coal began to fuel machines, trains and power stations.

The fossil-fuel boom transformed world history, switching major societies from an organic to an industrial metabolism. Before the use of coal became widespread, manufacturing and agriculture competed for two resources: land and labor. Manufacturing needed land to grow fuelwood and horse feed, and agriculture needed land for cultivation. Both needed laborers. But coal shed the ecological limits imposed by grown fuels, opening the door to constant growth.[13]

Coal was tremendously energy dense, with each kilogram containing three times more energy than a kilo of wood.[14] Productivity—what could be done with labor—soared. Great distances could now be traveled. Steamboats could ferry goods with speed. Machines could produce goods faster.

Industrialization was intensification. We were now able to cut trees faster, clear lands more quickly and pump greater volumes. Steam-powered plows dug deeper into the earth. New machines and vehicles accelerated the production, transportation and distribution of commodities. The barriers of space and time were eased.

The territory of the economy underwent major changes. Unlike water, coal could be stored, easily transported and accumulated. Previously water dictated industry, as productive cycles had to be arranged around it. But now industry could arrange its own path. No longer did factories need to be located near rivers, as they could be established anywhere where coal could be delivered.

Moving to coal power meant industrial production could be transplanted into urban centers, where workers were numerous. Industrialists in the 19th century turned away from water to coal power not simply because of coal's affordability or reliability, but because coal also allowed for easier control over labor. The urban poor were more easily disciplined than rural inhabitants.[15]

The Industrial Revolution was an unleashing of force, an explosion of more. More people, more products, more goods. The surpluses of fossil fuels handed industrialists an explosion of force equivalent to the use of slaves.[16] One barrel of oil roughly equates to 23,000 hours of human labor. Between 1850 and 1950, fossil fuels' share of total work output rose from 6.8 percent to 90.9 percent of work output, while animal labor's share decreased from 52.4 percent to 0.7 percent.[17]

The use of fossil fuels brought a huge burst of energy use, allowing for constant, regular and unrelenting growth. As historian Stephen Pyne notes, "coal, then petroleum and gas, were a kind of biotic bullion that acted on nature's economy like the plundered wealth of the Aztecs and Incas did on imperial Spain's."[18]

Energy wrought economic might. The first country to see major economic growth was The Netherlands in the 16th and 17th centuries, relying on domestic peat as well as Norwegian and Baltic timber. Britain then displaced Holland's imperial leadership through its vast reserves of coal. Later on, as the age of coal gave way to the era of oil and gas, the US took up the hegemonic reins.[19]

The injection of fossil energy transformed technology. Ships replaced their sails with coal-based steam engines, and then with oil-fuelled turbines. Street lighting shifted from tallow to candles, kerosene to paraffin, gas to electricity. Refrigerators replaced ice-boxes, cars replaced carriages, buses replaced trolleys. Communications were woven together through telegraph, telephone and radio systems. As car technology developed and the companies producing the new vehicles lobbied, cities were designed around the use of the automobile. The use of coal brought locomotives, and ignited the spread of railroads, the largest engineering undertakings of the 19th century. From Russia to India, hundreds of thousands of immigrant and indentured laborers set to clear space for tracks. Beyond mere technologies, these were enormous social systems laid over the land. In their needs for major organization and capital, railroads laid the foundation for the birth of a new institution: the modern corporation.[20]

In industrializing countries, the distances allowed by fossil fuels wrenched open wider gaps between urban and rural. Human beings, living in dense cities, were severed from nature. The consumption of food and energy was increasingly separated from its production. Hot water flowed magically from a tap, homes were heated in winter. As Dipesh Chakrabarty writes, "the mansion of modern freedoms stands on an ever-expanding base of fossil-fuel use."[21]

The pace of change was staggering. The forces of time and geology took millions of years to turn organic matter into fossil fuels. It has taken a portion of humanity mere centuries to extract and burn those fuels and convert them into powerful pollutants. By the 19th century, the use of fossil fuels had replaced deforestation and methane from rice cultivation as the principal source of human emissions.[22] And, as the fossil fuels burned in the early steam engines of industrialism, plumes of gas trailed into the atmosphere, smoke signals foreshadowing a future crisis.

Black gold: The story of petroleum

From its genesis, the stunning riches associated with oil allowed for the creation of massively powerful companies, extravagantly wealthy elites and rentier states. With much to be gained and lost, the history of oil has always been accompanied by corruption, conflict and controversy.[23]

In the early decades, when the oil industry faced public suspicion and political opposition through anti-monopoly trials, armies of lawyers were hired to fend off challenges. Advertising agencies were paid to plant articles complimentary to oil companies. Generous political contributions were deposited in the pockets of the powerful. The oil magnate John D. Rockefeller, in correspondence associated with a political donation to the Republican Party, wrote: "Our friends do feel that we have not received fair treatment from the Republican Party, but we expect better things in the future." In 1900, Standard Oil paid a Republican senator $44,500.[24]

Accounting tricks were used to concentrate powers while evading antitrust suspicion.[25] Companies built virtual monopolies, as they realized their profits could be maximized if they could control global markets and restrict supply. By the 1920s, a small selection of companies (including the precursors to Exxon Mobil and Shell) were in control of most major sites of oil production around the world.[26]

Industry and the creation of wealth had begun to depend on the use of oil. As oil wove its way into productive processes, demand soared. Major companies became empires. Between 1921 and 1929, Venezuela's production grew from 1.4 million annual barrels to 137 million barrels; half of its state revenue stemmed from oil. Between 1914 and 1945, Iran's oil exports rose from 300,000 tonnes to 16.5 million tonnes.[27]

Prospectors traversed the world, scouring for supplies. But the territories of oil largely overlapped with dense ecosystems and communities. The enormous Masjid-i-Suleiman concession in Iran overlapped the winter grazing pastures of the Bakhtiari peoples. In Venezuela, oil prospectors found swathes of oil strewn across indigenous territory. Shell fortified its own tractors to protect them from arrows from local communities.[28] For prospectors, these were obstacles that needed to be cleared. In Borneo, Chinese laborers were sent to cut through the jungle.[29]

The jungles of Carare Opun, in the Colombian region of Barrancabermeja, were inhabited by the Yariguí people. Little is known today about their existence. They were known to be fiercely brave, having resisted the incursions of outsiders for centuries.[30] But then, colonizers looking for tagua nuts in the outskirts of the region stumbled into fountains of oil. The Colombian state was quick to adjudicate the territory. Various oil companies would acquire the concession, before it fell into the hands of the International Petroleum Company, a Standard Oil subsidiary.

As the oil industry sought to settle in the region, the Colombian state gave resources to "reduce the Indians." In 1915, the governor of the Colombian department of Santander was reported as having approved a significant budget to fund a mission to "reduce and subject" the Indians.[31]

Metropolitan editorials portrayed the incursion of oil as the entrance of civilization into the jungle.[32] Yariguí communities were massacred, displaced or expelled. Children were forcibly placed in the hands of parishes or religious schools. Within years, the Yariguí people were entirely disappeared by the oil industry, wiped off the human map.[33] Hundreds of thousands of hectares of forest and mangroves were also cleared. This territory would become the largest foreign source of oil for Standard Oil, a company that endures today through its successors, Exxon Mobil and Chevron.

In another Colombian region, the land of Catatumbo, the state granted the Barco oil concession. Its territory coincided with the land inhabited by the Barí people, who had successfully resisted Spanish colonization and evangelizing missions. All measures were taken to make the area favorable for the oil companies. The concessionary contract had a clause which obliged the government to protect the contract companies by repelling the tribes of "savages." Armed groups attacked the Barí people to make way for drills. Many others died, electrocuted by the electric fences protecting oil installations. Within decades, they had lost two-thirds of their territory and half their population.

Historian Jacques Aprile-Gniset defined the mentality of oil extractors in the region: "before we perforate the soil, we have to 'perforate the inhabitants.'" Such genocidal tendencies accompanied oil's expansion across the globe. Petroleum was routinely placed before the right of communities to exist.[34] Oil industries penetrated into the depths of territory. Petroleum geologists located deeper deposits, engineers assembled even larger drillers. Forests were demolished to make way for facilities. Leaking pipelines were laid across the undergrowth, turning rivers into sewage. Displaced from their ancestral homes, the Sikuani, Betoyes, Hitnú, Hitanú and Dome Jiwi peoples went from being stewards of the Colombian savannah to begging in oil towns.[35]

In Nigeria, oil spills were endemic from the early 20th century, when the Nigeria Bitumen Corporation began its operations in Lekki Lagoon. Decades later, in 1956, enormous reserves of oil were discovered in the Niger Delta, primarily in the region of Ogoniland, home to the Ogoni people. Commercial production began in Oloibiri in 1958.[36] Since then hundreds of annual spills have taken place, leaking over nine million barrels of oil into the territory.[37]

In 1960, Nigeria gained independence, but the legislation and infrastructure of extraction remained largely in place, allowing oil companies to continue benefiting as they had in colonial times. Social movements emerged demanding sovereignty and an end to ecological atrocities.

In the region of Ogoniland, social movements organized mass protests. The backlash from the Nigerian police killed 1,000 people, displacing tens of thousands. In a notorious case, nine Ogoni men were executed. Despite widespread knowledge of the Nigerian military's abuses, Shell officials encouraged the suppression.[38] Amnesty International has accused Shell of "complicity in murder, rape and torture" in the region.[39]

In Ecuador, oil exploration began as the rubber economy disintegrated in the 1920s. The real rush for oil began in the 1960s, as companies sent drills and hundreds of thousands of workers into

Ecuador's Oriente, home to dozens of indigenous communities. Lands were sliced into extractive blocks.[40] Over the next three decades, hundreds of drills spilled billions of gallons of oil and toxic wastewater into soils and water streams. Texaco's activities destroyed the Tetete and Sansahuari peoples.[41]

In the early 1980s, Shell worked to extract gas fields in the Peruvian region of Camisea. Access roads were built, bringing in loggers and business interests. Around half of the Nahua people subsequently died following a series of epidemics after their contact.[42] Across the board, the oil industry established itself with little care or consultation, placing the worth of its combustive fuels before any other forms of worth.

Oil and power

Towards the end of World War One, French senator Henri Berenger remarked: "He who owns the oil shall rule the world."[43] Geology became fate, as countries accumulated enormous influence simply through access to reserves.

Countries and companies worked assiduously to secure diplomatic and economic agreements. States fostered climates hospitable to investment. The Venezuelan dictator General Juan Vicente Gómez told the US representative there in 1920, "You know about oil, you do the laws."[44]

All measures were taken to ensure the continuous flow of oil. In 1914, General Gómez asked Shell to build its refinery offshore, beside the island of Curaçao, taking advantage of its deep port but also fearing Venezuelan political turmoil.[45] Similarly, fearing the potential "radical tendencies" of future Venezuelan governments to "confiscate" oil properties, the Lago Oil company built its export refinery on the island of Aruba.[46]

Those that stepped out of line and expropriated oil companies were met with trade embargos. In 1951, Mohammad Mossadeq was elected prime minister of Iran. Upon his accession, he nationalized the country's oil fields, offering restitution to the oil companies. He was promptly deposed in a coup, backed by British and American oil interests.

Fossil fuels powered fierce geopolitical feuds and struggles for power. Lakes of blood were spilt in its name. Japan's attack on Pearl Harbor came as it sought to acquire oil fields in the East Indies.[47] Hitler saw the capture of the Caucasus' oil fields as crucial to his military campaigns. The black earth of Ukraine, and the black oil of Baku, were indispensable: "Unless we get the Baku oil, the war is lost." Oil interests contributed to many other wars, including the Chaco War in which over 50,000 Bolivians and Paraguayans died.

Following World War Two, the United States moved to secure global production and distribution of oil. Commenting on the Persian Gulf, a US official remarked: "The oil in this region is the single greatest prize in history."[48] In the wake of World War Two, the oil industry continued its explosive ascent as its center of gravity moved from the Caribbean to the Middle East.[49] Postwar diplomacy with Saudi Arabia, Kuwait, Iraq and Iran secured even greater elephants: the industry's term for giant oilfields.[50]

The scale of extraction boomed: larger refineries, larger drills, larger tankers, larger roads. The grand revenues of oil extraction gave governments the delusion of eternal bounty. Journalist Ryzsard Kapuscinki, travelling across Iran, observed that "oil creates the illusion of a completely changed life, life without work, life for free, it expresses the eternal human dream of wealth achieved through a lucky accident...it is a fairy tale and like all fairy tales a bit of a lie."[51]

With such riches for the taking, graft and corruption were widespread. From Turkmenistan to Saudi Arabia, fortunes were siphoned off through the theft of oil revenues. In the United States, before specific legislation was introduced in 1974, direct bribes by oil companies were common. One oil executive in Angola told journalist Ken Silverstein he spent 99 percent of his "waking hours trying to figure how not to technically violate the Foreign Corrupt Practices Act."[52] Oil also became synonymous with inequality. In Nigeria, 80 percent of national oil revenues flow into the pockets of just one percent of the population.[53]

And as companies and elites lined their pockets, fossil fuels grew increasingly central in the international economy. Between

1950 and 2000, global consumption of energy quintupled. Global GDP multiplied by seven, the human population doubled, and CO_2 emissions increased by a factor of five. Over the course of the 20th century, we experienced a 4-fold increase in world population, a 17-fold increase in the production of carbon dioxide and a 40-fold increase in industrial output.

Today, oil is not just a source of energy. It is a foundation of wealth, a raw source of power, the identity of institutions.[54] The great fortunes of the last century were effectively built around oil, in the shape of the automobile, petroleum, rubber, glass, steel, road-building, agribusiness, construction and real-estate industries.[55]

Decades of carbon capitalism have woven fossil fuels into the fabric of our world. Cultures have been shaped by their influence. Cars have dominated our streets and cities as well as our minds, emerging as symbols of prosperity, freedom and coming of age. As cultural theorist Imre Szeman explains, "Our expectations, our sensibility, our habits, our ways of being in the world, how we imagine ourselves in relation to nature, as well as in relation to one another—these have all been sculpted by and in relation to the massively expanded energies of the fossil-fuel era."[56]

The deceit and the delay

Fossil fuels are at the center of the most lucrative industry in history. For over a century, trillions of dollars have been made through controlling, extracting and trading reserves of oil, gas and coal. Any substantive resolutions of the climate crisis will directly compromise future profits of this kind.

Since its genesis, the oil industry has spent fortunes lobbying governments, acquiring political control and guaranteeing its relevance. As awareness of global warming and the need for collective action has arisen, oil companies have largely dedicated themselves to disseminating false science and to stalling a meaningful response to climate change.

Petroleum companies were advised as early as 1968 about the catastrophic climatic risks posed by their business. In July 1977,

Exxon's senior scientist James Black addressed the company's management committee: "In the first place, there is general scientific agreement that the most likely manner in which mankind is influencing the global climate is through carbon-dioxide release from the burning of fossil fuels."

Black warned that increasing emissions of carbon dioxide into the atmosphere could potentially increase global temperatures by three degrees. He urged the company to change, arguing that "man has a time window of 5 to 10 years before the need for hard decisions regarding changes in energy strategies might become critical."[57]

But, rather than heed the dangers, the company's executives doubled down. Just as corporations have funded campaigns to deny the dangers of tobacco, asbestos, lead, air pollution and plastics, fossil-fuel companies entered the business of denial, publicizing stories at odds with their own research. Exxon spent over $30 million funding think tanks which promoted climate denial.[58]

The goal was simple: spread doubt. As the American Petroleum Institute's action plan in 1998 explained: "Victory will be achieved when average citizens 'understand' uncertainties in climate science." This orchestrated confusion, and flagrant obfuscation of the uncertainties of atmospheric science, has helped popularize distrust around the science of climate change. Between 2005 and 2008, oil magnates the Koch brothers spent nearly $25 million lobbying against climate reform.[59]

In addition, companies worked assiduously to disrupt climate negotiations, derail environmental policies and prolong energy programmes favorable to their interests. Today, fossil-fuel companies receive huge backing from governments the world over in the form of production support, consumption subsidies, tax breaks, no taxation on externalities, loans, price controls and discounted land acquisitions. The International Monetary Fund has calculated that the industry had received $5.3 trillion in subsidies for 2016—more than the total spent on public health worldwide that same year.[60]

Our addiction to fossil fuels is, however, not some irrepressible outcome of history, or the unavoidable fate of technology. It is the outcome of historical maneuvers and choices. To build a coal and oil economy, huge infrastructure is required. This infrastructure was pushed for and implemented by a small portion of the world. Other options were always available.

Solar cells were invented in 1954, designed to power satellites in space. In the late 1970s, oil prices skyrocketed, sending state finances into disarray. Governments frantically researched solutions. Many countries poured huge amounts of money into alternatives to oil, spurring investments in wind and solar.[61] But, as the crisis eased, the alternatives were abandoned. As the activist Ralph Nader remarked: "The use of solar energy has not opened up because the oil industry does not own the sun."[62]

Recovering our historical memory

The book of memory is dense, and soaked in pain. To recall the plunderous and impoverishing effects of our history is to come to terms with a violent, degraded and forgotten past. But it is essential, for our amnesia, our mutilated memory, impairs our understanding. We struggle to comprehend the staggering present because we are unaware of our staggering past.

Recovering that historical memory is the first step towards acting upon it. By confronting the past, we can begin to address issues of restitution and reparation.

HUMAN NATURE
OR HUMAN IGNORANCE?

We are the land, we are in the land, in the land we are.
S Pushaina[1]

. . . .

Ambiguity is richness.
Jorge Luis Borges

. . . .

*In the old days we learned everything at once,
then we had to take it apart to understand it. When I went
to white school, I had to learn everything in little parts,
then try to put it together again. I thought that was backwards.*
Lakota elder, quoted by Kent Nerburn[2]

. . . .

*Our culture sets Nature as the highest bar for decorum,
while simultaneously giving Nature our lowest standard of respect.*
Alex Johnson[3]

. . . .

There will be no global justice without cognitive justice.
Boaventura de Sousa Santos[4]

. . . .

*They know their bodies are, in the most profound philosophical
sense and the strictest scientific sense, territory....
That the water of rivers also flows red through their veins.*
Horacio Machado Aráoz[5]

RECALLING OUR BRUTAL HISTORY of atrocities and ecological insensitivities provokes questions: how was such blatant violence possible?

Our mind reaches for explanations. The first story we might tell ourselves is that of human nature. We destroyed because we could not stop ourselves, or inhibit our unrestrainedly acquisitive and aggressive instincts. Humanity, the flawed species of boundless intelligence and cruelty, overrode boundaries, looting and polluting.

The second story might be that of innocence through ignorance. We destroyed because we did not know better. Only now, equipped with our contemporary knowledge and reason, do we have the tools to tackle the climate crisis.

These two positions, in their variations, underlie a lot of popular environmental thought. Appraising their validity is necessary, for our diagnosis evokes our prognosis. The assessments we make lay down the road maps ahead.

What human nature?

To be human is to be able to shape the environments we inhabit. But to be human is not to destroy. Just as we have eroded the ecosystems we rely on, human beings have also enhanced and restored them. The history of nature is one of both destruction and defence, protection and plunder, attention and carelessness.

When we see vast stretches of rainforest, we often imagine pristine areas, unspoiled by the damaging touch of humans. Many of us have grown up with the myth of separation, taught to see safe nature as empty of human life. But forests have always been inhabited by human beings. What we imagine as virgin forests are often cultural landscapes. The vast majority of our planet's territory has been inhabited and marked by humans across their history.[6]

The inhabitants of these lands held diverse visions of nature, some rooted in distance, but most in devotion. Forests, seas and all spaces of life were populated by forces outside of human control, the venerated homes of spirits and gods.[7] Many peoples saw and continue to see themselves as inseparable from the rivers, trees, air currents, rains and animals that surround them. This nonhuman world was imbued with language, spirit and soul. Life evolves as dialogue, mutual and reciprocal connection with other beings.

In Papua New Guinea, Gimi villagers have long understood the proximity of people and forests, believing that the life force (*kore*) of humans who are buried returns to and brings life to the *kore* of the forest.[8] In the Ituri forest of the Congo, Mbuti communities refer to their surrounding forest as a parent, "Father" or "Mother." The forest is seen as a source of food, shelter, clothing, warmth and affection.[9] In Wayuu communities of Colombia and Venezuela, the earth is regarded as a body, or an extension of the human body. The Yoruba of Nigeria speak of "Ayé," a physical realm of entangled beings and events.[10]

The ecosystems around us are infused in our founding mythologies. Many of our stories of creation locate our origin in the sea, in the sky, in maize or in the land. In Biblical terms, Adam, the Hebrew name for the first man, comes from *adama*, earth or soil. Eve stems from *hava*, the Hebrew for living.[11]

For many societies, nature was the closest route to divinity. In areas of medieval Europe, nature was widely seen as a record of God's will.[12] For the rabbinical scholar Maimonides, the study of the natural world was a prerequisite for the study of the spiritual world, because it taught awe and humility.

Nature suffused the cultures, languages and peoples it nourished. In ancient Scotland, the older Gaelic alphabet had letters made of twigs. Each letter corresponded to a tree. The first letter was a birch tree, used to begin or ignite fires. The final letter of the alphabet was a yew. Yew trees, planted in cemeteries, were associated with ending.[13]

Such intimate connections led many societies to assert the importance of care for our life-giving surroundings. Institutions and social norms were developed to manage practices. Self-destruction was controlled through taboo. Everyday rituals and ceremonies celebrated and offered gratitude to the abundance of life. Children, from an early age, were instructed in the library of life that surrounded them, learning ancestral stories about local flora, fauna and topography.[14]

These practices of coexistence and dependence generated mutually beneficial relationships between human beings and their environments. Various studies have shown how the biological diversity of our habitats has been promoted and developed by indigenous and peasant peoples.[15] The Amazon was inhabited by thousands of communities who actively enriched the forest. In the region of British Columbia, First Nations communities nourished coastline forests, improving their fertility and soil quality.[16] In Guinea's Kissidougou region, local farmers have historically transformed and replenished the landscape, planting protective forests around human settlements.[17] European peasant economies, until the 19th century, tended to maintain high levels of biodiversity.[18]

In this sense, human beings have been vital participants and contributors within ecosystems, honing a mutually supportive relationship between culture and biology. From Papua New Guinea to Brazil, Mexico to the Democratic Republic of Congo, it is no coincidence the areas of the world with the strongest cultural and linguistic diversity overlap with the areas of strongest biological and agronomic diversity.[19]

As a general rule, human diversity concurs with environmental diversity. Although indigenous territories make up just a fifth of the Earth's surface, they hold four-fifths of the world's biodiversity.[20] As the Kuna leader Geodiso Castillo notes, "where there are forests there are indigenous people, and where there are indigenous people there are forests."[21] The geographer Bernard Nietschmann called this "symbiotic conservation." As Nietschmann noted:

The vast majority of the world's biological diversity is not found in gene banks, zoos, national parks, or protected areas. Most biological diversity is in landscapes and seascapes inhabited and used by local peoples, mostly indigenous, whose great collective accomplishment is to have conserved the great variety of remaining life forms, using culture, the most powerful and valuable human resources, to do so.[22]

But this intimate connection of interdependence also means that threats to biodiversity threaten cultural diversity. The extinction of languages mirrors the extinction of species.[23] As we destroy eco-systems, we are also torching the life-worlds humans have helped sustain.

The myth of collapse

Despite the ubiquitous ability of humans to nourish life in their environments over history, these stories are frequently obscured by narratives of collapse and ecological ruin. According to this vision, powerful civilizations, from the Khmer to the Maya civilization, collapsed through the stupidity of unrestrained resource use. Deforestation, unsustainable urbanism, degrading agriculture and silt-accumulating irrigation all led to the downfall of magnificent states.

The story of collapse, and its depictions of chaos and disarray, is often invoked to exemplify the power of human delusion and foolishness in nature.[24] But while it may make for captivating reading, it is widely overstated.

First, history is riddled with convoluted visions of what a civilization is. Civilizations ebbed and flowed, rose and declined, changing form. Egyptian civilizations did not vanish, but rather morphed, influencing Hellenistic and Roman civilizations. As such, framings of collapse are unclear, riven by value judgements.[25]

The Mayan "civilization" was an interconnected assemblage of dozens of cities and states. Rather than drastically disappearing, the civilization transformed. Populations in some areas fell, the rule of certain leaders was rejected, and some cities were abandoned. When the Spanish conquered the area in the 16th century, those communities were still there. Although the Mayan kingdoms were destroyed by the Spanish armies, millions of Mayan descendants and dozens of communities endure across Central America.

Across many case studies, the evidence for mass delusion is far weaker than suggested. Easter Island is another emblematic example suggested to prove the folly of human stupidity. According to this interpretation, Easter Islanders felled all their trees, failed to control rampant population growth and provoked their own war-fuelled doom.

But contrary to the traditional story of environmental collapse, the island's local populations appear to have been inventive farmers, managing limited resources quite sustainably.[26] But they were then targeted by slave raiders, who systematically kidnapped inhabitants and compelled them to work on guano mines in the Chincha islands.[27] British sailor HV Barclay observed in 1868: "It is a sad fact that in these [Easter] islands, as in North America, wherever the white man establishes himself the aborigines perish."

Collapse seen as total extinction quickly takes us down conceptual dead ends. There are, however, warnings to heed. As some societies grew, and became more complex, they often tended to get more difficult to manage. The anthropologist Joseph Tainter defined collapse as the rapid loss of complexity in a society. Complexity, in his eyes, meant more territories, roles, activities and institutions. Tainter saw that, across history, the complexities of societies would rapidly increase and diminish. Sustaining complex systems requires enormous influxes of energy. Early states would deploy agriculture or the consumption of forests, while modern states require fossil fuels.

As the challenges of complexity grew, many societies adopted short-termist approaches. But, over time, returns begin to diminish. Farming land becomes less productive, forests are depleted, and mines are emptied. Extravagant elites drain diminishing resources. The cost of managing over-stretched empires and bloated administrations booms. In such conditions of fragility, any climatic shock or military invasion becomes enough to topple regimes.[28]

Although any grand theory such as Tainter's misses local specificities, scholars are more concerned with the modern applicability of its warnings. Our software, machinery, legal systems and economies are gaining unprecedented levels of complexity.[29] Our societies are increasingly interconnected, yet the problems they face are ever more entangled and illegible.[30]

While there are reasons to be concerned about human decision-making, particularly in the face of complexity, we should always look askance at any invocation of human nature, for there is no single one. History's exuberant diversity shows us the amplitude of human behavior, and the sheer range of creative social configurations societies have experimented with.

Our fixation on human nature easily becomes an argument for indifference. If our nature is to blame, there is no exit. If history is a process of doomed civilizations that rise and collapse, we are destined to repeat it. Destruction is an immutable, natural law. Recalling our possibilities, and the complications of the past, offers us other ideas. Perhaps, we have more of the answers and abilities than we might like to acknowledge.

A history of knowledge and ignorance

Human beings have always sought to understand their surroundings. The penchant for questions and the willingness to experiment are not inventions of the European university. For millennia, we have been scientists: answering questions with questions, testing our ideas against the world around us. Through observation, intuition, inference, interaction, experimentation, verification and scrutiny, we sought insight in the laboratories of life.[31]

We learnt to understand the colors of the sky, interpret the shape of clouds, read the territory we walked and sense the texture of sands. We mapped the web of life, distinguishing hundreds of species, dozens of shades of snow and diverse types of soil.[32] We learned about fertility and how to endure droughts and storms. [33]

Through intimacy with the rhythm of nature, communities learned to predict weather patterns, building sophisticated understandings of rain corridors, storm routes and wind rhythms.[34] Polynesian societies navigated the vastness of the Pacific Ocean, a quarter of the planet's surface, by reading the skies, stars and movements of marine animals.

Existence and flourishing relied on this knowledge, passed on through oral stories. For hundreds of generations, across thousands of years, in thousands of societies, humans refined their ability to nourish life in their environments. Ours is a history of weavers, hunters, engineers, hydrologists, architects, agronomists and fisherfolk, living and co-evolving with their surroundings, crafting a creative memory of survival.

Together, we have assembled a tremendously rich biocultural memory, a wealth of teachings and insights into land use, medicine and climatology. Human communities developed immense agronomic expertise in their own environments, learning to grow their own variations of crops that could thrive and enrich the biological diversity of their broader ecosystems.[35] Agricultural practices, from terraces to composting to crop rotation, were adapted to recycle nutrients, prevent erosion and boost fertility.

The Moru peoples of Sudan, for example, have historically used expert soil classification, organic waste application and the flexible rotation of crops to preserve fertility. In the Himalayas, Chepang communities practise *khoriya*, rotating crops and leaving fields fallow for years on end. Mesoamerican societies practise *milpa*, an agroecological system shifting the cultivation of corn, beans and squash. In various sub-Saharan societies, communities nurtured dark earth, exceptionally nutrient-rich soils, by mixing ash, biomass, char, bones and organic waste.[36] In Burkina Faso, Mossi communities have used innovative concentric ring circle formations to enrich fertility.[37]

Various societies created forest gardens, introducing herbs and shrubs into forests. The coexistence of edible plants and endemic biomes has helped maintain the vitality and structure of forests. In Tanzania, Chagga communities use the *kihamba* system, where overlapping layers of vegetations interact. Trees provide medicine, firewood and fruits; under the trees, lower-hanging fruits are grown, while beneath these, shrubs and vegetables are planted. In northeastern India, Apatani communities developed cultivation practices (*jhum*) involving advanced rice cultivation through gardens which imitate forests. The *shamba* (vegetable farms) of Uganda, the *kebun-talun* (human-made forests) of western Java, the *pekarangan* (home gardens) of Sumatra, the *kandy* of Sri Lanka and the *Kuajtikiloyan* of the Mexico's *Nahua* peoples are all vivid examples of this tradition.[38]

Some of the most effective techniques deployed to prevent drought or flooding are rooted in ancient traditions. In Chad's Sahelian areas, farmers have used rainwater-harvesting techniques (Zaï), which involve cultivating crops in small pits. In India, *johads*—traditional crescent-shaped dams made from earth and rocks—are used to collect rainwater and replenish water tables. In Afghanistan, communities afflicted by flooding developed ingenious early-warning systems, whereby *mirabs* (water masters) are tasked with warning downstream villages of impending floods.[39]

Inventive technologies adapted agriculture across diverse environments. Terrace farming was developed thousands of years ago in Ethiopia, Mesoamerica, China and Korea. Around Lake Titicaca, in modern Peru and Bolivia, a hydro-agricultural system called *waru-waru* covered hundreds of thousands of hectares.[40] In the valleys of Mexico, ancient inhabitants created *chinampas*, floating gardens with complex agrohydraulic systems, that stretched over thousands of kilometers. In Bangladesh, communities have honed the use of *baira*, floating plots of land made out of hyacinths; these mats float on water, absorbing nutrients and enriching biodiversity by offering homes for both aquatic and bird life. In Algeria, farmers use the *ghout* system, which plants date palms to create oases. In the northern Americas, indigenous peoples such as the Kwakwa-ka'wakw built clam gardens, terraced rock walls that provided homes for edible shellfish.[41]

Institutions and discussions

Maintaining these practices required significant effort, knowledge and coordination. Across many societies, collective institutions regulated and sustainably managed common resources—forests, waters, fisheries, pastures. Since the ninth century, farmers in the rice fields of Bali have practised *subak*: a sustainable method for collective water management rooted in sharing. Across Fiji, communities arranged *tabus*, protected fishing areas that were declared off limits for months on end. In Romania, there were forest commons called *obstea* (togetherness).

Reliance on nature generates extensive debate about how best to thrive within it. In this sense, ecological discussions, often rooted in concerns around damage to ecosystems, have been with us since our origin. Communities were immersed in conversations on ecological problems, in the form of inundation, buildup of silt, salinization, crop failure, drought, pollution, waste disposal and wood shortages.

In ancient China, Taoist and Confucian schools of thought debated the merits and dangers of hard and soft interventions in rivers. Many engineers realized that rivers could not simply be trampled or built over; paths for water had to be opened to avoid major flooding. Nature had to be adapted to, not commanded. The Taoist principle of wu wei (non-action or non-doing) was repeatedly applied to the management of water systems. Chia Jan, a Taoist hydraulic engineer, argued that the Yangtze River should be given plenty of room to take whatever course it wanted.[42]

Centuries later, in 1786, the parliament of the French city of Bordeaux spoke on behalf of nature to dissuade the central government from pursuing a project to regulate the Dordogne River. They argued that the river maintained a "perfect balance which commanded even reason to silence."[43]

Philosophers and state officials warned of the corrosive effects of environmental destruction. In the year 200, the author Tertullian warned of the impacts of Roman expansion: "All places are now accessible, well known, open to commerce. Delightful farms have now blotted out every trace of the dreadful wastes; cultivated fields have overcome woods.... We overcrowd the world. The elements can hardly support us. Our wants increase and our demands are keener, while Nature cannot bear us."[44] Pliny the Elder, watching the emergence of urban landlords and slave-driven cash crop agriculture, lamented that such practices would ruin the empire.[45]

In China, fourth-century philosopher Pao Ching-yen argued "that the bark is peeled off the cinnamon tree, that sap is collected from the mountain pine, is not what the trees want. That pheasants are plucked and the kingfisher torn to pieces is not the wish of these birds.... The seeds of fraud and cunning lie in acting against nature, when such an action is based on violence."[46]

With the advent of colonialism, ecological arguments proliferated across colonies and colonial states, arguing for the need to protect environments and modify despoiling and profligate practices of economic development.[47] Many colonized communities were aware of the dangers of rapaciousness, with elders warning of the destructive unviability of settler societies based on endless conquest.[48]

Concerns around excessive pollution have similarly been abundant for thousands of years. Many of the earliest scholars observed the deleterious consequences of activities such as mining.[49] Given abundant fears of disease, words like "environment" were synonymous with health, and quality of life was associated by many with the quality of the environment.[50] Hippocrates, the forefather of medicine, argued that if you wanted to determine the health of a person, you had to look for where their home was located.[51]

Gradually, the notion of limits on resources began to be introduced. In 1556, metallurgist Georgius Agricola, known as one of the early theorists of geology, recognized the dangers of dependence on nonrenewable resources: "the tunnels eventually cease to give metals, while the fields always yield crops."[52] In 1827, two miners from the German Ruhr region issued a memorandum calling for an end to intensive coal extraction, "since everything has its limits."[53]

Laws and actions

These were not just rhetorical debates. Early societies were alert to the importance of sustainable resource management, arranging laws to steward resources and protect common environments. The Babylonian Code of Hammurabi, one of the oldest known written codes, speaks extensively about irrigation practice and water protection. One of the ancient Indian Emperor Ashoka's edicts outlawed the needless killing of animals and the torching of forests.[54] The Sumerian kingdom of Mari introduced measures to protect mountain forests that were being rapidly cleared. In 450 BCE, King

Artaxerxes I of Persia called for the cutting of cedars in Lebanon to be restricted.[55]

Since antiquity, sacred groves and parks dedicated to deities were planted and protected. In the Zoroastrian tradition, lush gardens were cultivated to imitate the celestial plain; the word "paradise" itself comes from the Avestan word *pairidaeza*, referring to these Persian gardens.[56]

From Sindh to Tokugawa, Venice to Zurich, states and city councils throughout the ages introduced legislation controlling fishing practices, protecting forests, banning polluting activities, regulating extraction, encouraging the plantation of seedlings, outlawing the hunting of particular species and recycling excrement as fertilizer.[57]

And when power contravened on the ecological relationships communities relied on, resistance was ubiquitous. In the 18th century, the Maharaja of Jodhpur wanted to build a palace using wood from a forest in Khejarli. A group of Bishnoi Hindus reliant on the forest resisted, protecting the trees. Hundreds were slaughtered.[58] In 1760, widespread resistance erupted in the Chinese region of Hangzhou in opposition to a quarry, arguing that it disrupted the natural dynamics of the Earth.[59] In 1765, citizens in the German city of Cologne mobilized and closed a lead foundry, articulating in their demands that "they should not be driven from their homes or poisoned in them."[60]

As industrialization accelerated, many workers pressed for better working conditions and legislation on pollution. Environmental historians have unearthed swathes of concerns, protests and uproar during the industrial era. In Paris, in the early 19th century, there were accounts of constant battles between neighbors and factories.[61] As sociologist Barrington Moore concluded, "there is no evidence that the mass of the population anywhere has wanted an industrial society, and plenty of evidence that they have not."[62]

In 1890, inhabitants of Japan's Watarase River valleys protested the poisoning of their rivers by the Ashio copper mine. Village leader Shozo Tanaka became an outspoken environmentalist, leading major social movements that pushed for Japan's first law addressing industrial pollution. Criticizing the development pursued by Meiji oligarchs, Tanaka advocated care for "mountains, forest and rivers," arguing that "to destroy the village is to destroy the country."[63]

Climate change and human influence

For millennia, human beings were aware that their actions could bear profound consequences on the climate and ecosystems around them. The relationships between forest removal, droughts and erratic rainfall were widely acknowledged.[64]

The Greek naturalist Theophrastus observed a lake near Larisa (Thessaly), and began to speculate on how human transformation of the landscape, through draining marshlands and felling woodland, could affect temperature (making the climate colder).[65] In the 16th century, the scholar Yen Sheng-Fang recognized the link between deforestation, soil exhaustion and water depletion.[66]

In Africa and the Americas, early colonizers knew of links between deforestation and environmental change. Christopher Columbus, aware of the experiences of the Canary Islands and the Azores, knew that forest removal would decrease soil moisture.[67]

But these insights did little to stop intensively destructive plantations. Only after centuries of agrarian malpractice did various colonial officials in the Caribbean act to restrain exhaustive farming practices and deforestation. In 1791, laws were passed on the Caribbean island of St. Vincent, protecting forests to preserve rainfall. The Atlantic island of St. Helena became the center of experiments with conservation and the encouragement of rainfall.[68]

Around this period, naturalists increasingly began to consider the broader influence of human beings in nature. In 1844, German scientist Alexander von Humboldt determined that human beings were affecting the climate, "[t]hrough the destructions of forests, through the distribution of water, and through the production of great masses of steam and gas at the industrial centers."[69] Concerns surrounding human activity's effects on the climate were widespread throughout the 19th century; historians of the era were acutely aware of the links between forest clearance, aridity and flooding.[70]

Scientists studying fossils and geology shed light onto the fragility of species, including ours.[71] French naturalist Georges Cuvier researched dinosaurs, theorizing that a major catastrophe could have wiped them out. Cuvier's assistant, Louis Agassiz, pioneered understanding of the Ice Age.

In 1824, French physicist Jean-Baptiste Fourier began describing the Earth's atmosphere as an insulator, trapping heat like a pane of glass.[72] Three decades later, Irish physicist John Tyndall performed experiments that proved the heat-trapping ability of carbon dioxide. In 1896, Swedish physicist Svante Arrhenius theorized that burning coal was heating the planet. His calculations predicted that temperatures, particularly in the Arctic, would soar if the amount of atmospheric carbon doubled.[73]

The Russian geochemist Vladimir Vernadsky, who came up with the idea of the biosphere, stressed the growing influence of humans on the Earth's biogeochemical cycles.[74] Polymath Pyotr Kropotkin challenged Victorian ideas of the climate as a stationary environment, seeing climate change as a force within history.[75]

In 1917, scientist Alexander Graham Bell wrote that unlimited burning of fossil fuels "would have a sort of greenhouse effect." "The net result is the greenhouse becomes a sort of hothouse," he added. These ideas began to lay solid foundations for our modern understanding.

In 1959, as the US oil industry celebrated its centenary, nuclear physicist Edward Teller addressed a symposium attended by public officials, academics and oil executives:

> I believe that the energy resources of the past must be supplemented.... Whenever you burn conventional fuel, you create carbon dioxide.... Carbon dioxide has a strange property. It transmits visible light but it absorbs the infrared radiation which is emitted from the earth. Its presence in the

atmosphere causes a greenhouse effect...a temperature rise
corresponding to a 10 percent increase in carbon dioxide will
be sufficient to melt the icecap and submerge New York. All
the coastal cities would be covered....I think that this chemical contamination is more serious than most people tend to
believe.[76]

The secrets of the future lay in the past.[77] Throughout the late 20th
century, scientists looked more closely at the evidence locked in our
ice sheets, rocks, trees, sediments and fossils. By examining archi-
val documents, tree rings, ice cores, pollens, rock chemistry analyses and lakebed sediments, researchers assembled more detailed
pictures of the planet's natural history. Scarcely known fields such
as dendrochronology, palynology and glaciology proved essential
to the charting of our Earth's climatic antecedents.

The arithmetic power of supercomputers helped climate scientists craft even more sophisticated models. The idea that human
activity was heating the planet became unassailable. In 1975, influ-
enced by meteorologist Bert Bolin, the Swedish government drafted
a future energy policy: "It is likely that climatic concerns will limit
the burning of fossil fuels rather than the size of the natural resources." In 1977, the US National Academy of Science was warning
that unrestricted emissions would raise temperatures by up to 5.5°C,
and inflate sea levels by six meters.[78] In the 1980s, as sweltering heat
and intense fires afflicted the US, NASA scientist James Hansen delivered a historic testimony at the US Senate: "The greenhouse effect has been detected and it is changing our climate now."[79]

Voluminous pages began to be filled explaining the relationship between human activity and climatic changes, clearly laying
out the threat posed by rising greenhouse-gas emissions. In 1988,
the Intergovernmental Panel on Climate Change (IPCC) was established, releasing its first report two years later.

In 1992, over 1,700 of the world's foremost scientists signed an
open letter titled the "World Scientists' Warning to Humanity." It
cautioned:

Human beings and the natural world are on a collision course. Human activities inflict harsh and often irreversible damage on the environment and on critical resources. If not checked, many of our current practices put at risk the future we wish for human society and the plant and animal kingdom, and may so alter the living world that it will be unable to sustain life in the manner that we know. Fundamental changes are urgent if we are to avoid the collision our present course will bring.[80]

The letter went on to call for a movement "away from fossil fuels to more benign, inexhaustible energy," and a new ethic of responsibility for the Earth. But such exhortations have not necessarily translated into effectual action. As climate scientist Kevin Anderson laments:

In the quarter of a century since the first IPCC report we have achieved nothing of any significant merit relative to the scale of the climate challenge. All we have to show for our ongoing oratory is a burgeoning industry of bureaucrats, well-meaning NGOs, academics and naysayers who collectively have overseen a [60 percent] rise in global emissions.[81]

The impotence of knowledge

Examining the history of ecological knowledge displays a worrying disconnect between what we have known and how societies have behaved. Our contemporary reality exhibits this glaring paradox: we have more data, evidence and arguments for ecological action than ever before. Yet the level of political responsiveness bears little relation to the density of available information.[82]

What explains this? There are plenty of answers that can be offered to this question, but one important and often neglected area lies in the history of science, and its relationship to power.

Rather than a simple story of innocence through ignorance, the history of knowledge suggests something deeper. Among the roots of our ecological crisis sits the ability of the powerful to ignore warnings, trample on viable traditions, deploy skewed reasoning to sanction destructive behavior and misroute intellectual energies to intensify the problem. In many diverse situations, societies immersed in the ambitions of empire-making, and enraptured in the violent pursuit of wealth, neglected long-term decisions, and eventually found themselves unable to respond to dangers, especially chronic ones that accumulated over time.[83]

Contemporary states and institutions continue to disregard, deny or underestimate ecological problems.[84] Prophetic warnings continue to fall on deaf ears, as science, envisioned as a terrain of knowledge separate from nature and insulated from social values, is applied to accelerate the process of destruction.

Today, we find ourselves in a situation where scientific knowledge, the mast of modernity, is fragile. In the purported age of "post-truth" and "alternative facts," faith in experts has weakened. The doctrines of climate denialism and ecological optimism have done much to sow doubts around the truthfulness or relevance of climate violence.

In such a context of mistrust, it is tempting to preach for greater acceptance of science and wider dissemination of knowledge. But this is insufficient; we don't need only to defend science, but to transform it.

In part, we must recognize that certain visions of scientific knowledge have also contributed to the crisis we find ourselves in. The ideas that have brought us to where we are today will not be those that lead us out.

To chart a path ahead, we must come to greater terms with complexity and diversity, looking critically at some of the legacies of modern science, and crafting a model of knowledge equipped for the convulsions of the future ahead.

Science as a way of thinking

Science is not a catalogue of frozen truths, or a self-correcting factory of clarity. Rather it is an evolving process of learning, an endless attempt to find clearer pictures of the world around us.[85] In this eternal endeavor, what we see as true is profoundly provisional.

The history of knowledge is one of revolutions, where unquestioned assumptions are burst, inescapable paradigms are disrupted and heresies become truisms.[86]

While science strives to find answers approaching truth, it will irredeemably be a human effort, conceived of and performed by humans. All research originates somewhere, in a particular social context. Values, existing paradigms and cultural perceptions shape our queries, our approaches, our norms and our forms of reasoning. Inevitable biases bleed into the questions we form and the answers we seek. No scientist, however wedded to the allure of objectivity, can be a perfectly removed observer. Our context always affects what we take for granted, the way we think when we don't think we are.

As anthropologist Richard Nelson reflects:

> Reality is not the world as it is perceived directly by the senses; reality is the world as it is perceived by the mind through the medium of the senses... reality in nature is not just what we see, but what we have learned to see.[87]

The way we understand and approach life—our own and that of our surroundings—varies tremendously across the world.[88] As people, we organize reality along the lines of our perception. To be is to see but also to be inevitably oblivious to other meanings and views. Our prisms are partial, riddled with unconscious blindspots. Culture shapes our cognition, affecting what we notice, what we can describe, where our attention goes, where we look, what we overlook: the mental maps we make of the world.

Do we see dreams as hallucinated inventions, or reality in another form? Is nature a subject or an object? Are animals an "it" or "they," a fellow being or an inferior creature? Do we see human beings as creatures infused by spirit or as complex machines? Do we conceive of the brain as a computer, a storage of representations, a bank of memory or a dense interaction?[89]

Our grammar and vocabulary further delimit the world we can describe. Some languages have no hierarchies between the subject and the object, meaning all verb constructions are reflexive: "I walk on the earth that holds me" or "I breathe the air that breathes me." There is no action without reaction. In some languages, the words for "plant" or "land" are not singular nouns, but phrases translating to "those who care for us."[90] While many languages place major emphasis on "I," some languages have no possessive pronouns, their words constantly invoking and involving a "we."[91] Many languages or worldviews do not include the idea of nature. How can you name what you are part of?

What these examples show us is that assumptions about reality are always present in any scientific process or method. We always must impose language, space, time and parameters on a world that confounds them. In more practical terms, what is studied or understudied will depend on the questions we prioritize and the resources we allocate. Even researchers conducting seemingly dispassionate quantitative studies must choose what effect size is chosen, how variances are calculated and how statistical models are built.

Surveying the history of science shows just how vulnerable knowledge production is to its unacknowledged broader context. Although the name of science is brightened by countless valuable insights, it is also besmirched by its manipulations. Under the canopy of objectivity, scientific methods of reasoning were applied in many cases to validate hierarchy and exploitation. Evidence was cherry-picked and fitted into presuppositions. Edifice was assembled to solidify artifice.

Specious sciences, such as phrenology and craniometry, were used to rationalize racism. Statistics and the idealization of averages contributed to eugenics, the creation of skewed ideas of "normality" and overstated assumptions of "above average" and "below average."[92] In many states, baseless psychiatry was deployed to pathologize dissidents and protesters as mentally ill.[93]

Pervaded by prejudice, studies of gender in biology, neuroscience and psychology helped to legitimize the supposed inferiority of women. Deeply rooted stereotypes, positioning men as the stronger sex of traditional hunters and women as nurturing selves suited to caring, were rooted in dubious science. Reductionist biological studies looked for conclusive sex differences, neglecting the influence of culture or patriarchal structure. Individual variation was erased through innate gender variation. [94]

Similarly, the study of nature was used in ways that could legitimize what certain thinkers considered to be "moral," "normal" or "natural": "moral relations," "normal families" and "natural sexualities." Our depictions of nature continue to be suffused by idioms— such as the "the survival of the fittest," or evolution as a quest for supremacy—that often are narrow human projections of the world onto nature.

Science, for all its aspirations toward impartial isolation, has never been removed from questions of ethics. Those humans deemed disposable by the powerful were acceptable for research. Slaves, prostitutes, mental patients, prisoners and racially oppressed groups were subjected to grossly unethical experiments, left untreated, exposed to hypothermia and injected with cancerous cells or diseases.[95]

In the colonial era, imperial states placed great importance on the acquisition of knowledge. When explorers and colonizers set out, one of their first tasks was to collect and understand the natures of the "new" terrains. Technical experts and cartographers were dispatched to map the territory. Botanists and agronomists identified the most productive, high-yielding variations of plants. Leading scientific bodies extracted specimens from colonial territories, in collaboration with imperial companies. "Exotic" human beings were forcibly removed from their homes to be exhibited in museum displays.[96]

As they traveled, colonial scientists categorized the "natures" and "abnormal behaviors" they encountered. As Linda Tuhiwai Smith notes: "hierarchical typologies of humanity and systems of representation were fuelled by new discoveries."[97] Race was devised as a potent tool of distinction and categorization. Just as scientists could order the animal kingdom into species, human beings could be classified into biological races. Under this logic, races were distinct peoples, distinguished by physical phenotypes, with separate cultures, values, demeanors and intellectual abilities. Peoplehood was tied with biology. This fiction, under the imprimatur of supposedly sound science, would bear devastating material consequences.

The science of exploration coincided with the science of exploitation, as colonial thinkers expended their energies conceiving of how to better control and use nature.[98] In India, Scottish botanists William Roxburgh and Francis Buchanan were influential in shaping colonial forestry policies. Roxburgh recommended the clearance of "our impenetrable forests which cover very large tracts of the best lands in India." Buchanan called for state control to be exerted over the forests of Malabar so as to end their "deification."[99]

Science was the tool of dominion and distinction. As time has passed, social movements and thinkers have pushed researchers to question frames and impulses previously seen as unquestionable. Yet there is still a long road to go.

Science's contemporary challenges

The social context for knowledge today is engulfed in challenges. Many scientists have themselves diagnosed a triple crisis in science: a crisis of public trust in science, a crisis in the quality and credibility of published science and a crisis in the political use of science.[100]

The exponential rate of publication and the overflow of data have stoked fears of scientific saturation.[101] The entrenchment of hyper-specialization in academia has been flagged as a major impediment to the interdisciplinary conversations required to address crosscutting social issues.[102]

The privatization of research has also sowed concerns. Starting in the 1980s, corporations began to displace national governments as the main funders of research. Universities began to be contracted to undertake studies, and science increasingly became a commodity, sold for a price.[103]

Today, the pursuit of knowledge is increasingly colored by commercial and political pressures, which shape agendas and priorities. What gets studied is often a matter more of economic interest than of public importance or humanitarian priority.[104] Some lives are deemed worthier than others. Diseases that primarily afflict those with emptier pockets are less relevant.

Under this model of science, knowledge becomes a private good, often only accessible to those who can buy it, and only visible when promoted by those with the capacity to advertise it.[105]

Research, and its guiding assumptions, is molded or retrofitted to fit preexisting interests. Such institutional pressures tend to generate policy-based evidence, rather than evidence-based policy.[106] In its worst expressions, this has led to downright intellectual corruption.[107] In 1978, two academics, Bruce Owen and Ronald Brauetigam, advised business leaders: "Regulatory policy is increasingly made with the participation of experts, especially academics. A regulated firm or industry should be prepared whenever possible to co-opt these experts. This is most effectively done by identifying the leading expert in each relevant field and hiring them as consultants or advisors."[108] Following their advice, industry money has poured toward funding favorable studies, influencing journalists, bankrolling conferences and popularizing conclusions.

Over recent decades, social movements have had to confront the "scientific evidence" marshaled by the tobacco lobby, the asbestos lobby, the pesticide lobby and the diesel lobby. Historians Naomi Oreskes and Erik Conway have rigorously documented how corporations have for decades hired "merchants of doubt," designing public-relations schemes to distort and obfuscate the science. Although companies knew internally about the dangers of their products, they organized campaigns of deceit. Scientists and lawyers were contracted for lobby campaigns specializing in the stoking of uncertainty. Industry regulators and scientific institutes were co-opted.[109] In the United States, between 2003 and 2010, over $558 million were spent on climate misinformation campaigns.[110]

As these trends took off, and qualified scientists faced increasing work burdens, the wider quality of science began to diminish. Today, science is facing a major crisis of reproducibility, as evidence accumulates suggesting that significant portions of the scientific literature, from medicine to economics, forensic science to organic chemistry, may be misleading.

One key marker of the strength of a scientific study is its reproducibility: the ability of other researchers to replicate a study and reproduce its results. A study that is not replicable does not mean the study is false, or that its entire angle is erroneous, but it does raise major questions about its truthfulness.

The researcher John Ioannadis, who has studied reproducibility, expressed his concern that "in modern research, false findings may be the majority or even the vast majority of published research claims."[111] Richard Horton, physician and editor of *The Lancet*, added:

> The case against science is straightforward: much of the scientific literature, perhaps half, may simply be untrue. Afflicted by studies with small sample sizes, tiny effects, invalid exploratory analyses, and flagrant conflicts of interest, together with an obsession for pursuing fashionable trends of dubious importance, science has taken a turn towards darkness.[112]

Over half of preclinical biomedical research and published psychological research has been found to be irreproducible.[113] Reviews of research into strokes, and the most vital cancer studies, found only 15 percent and 11 percent respectively to be plausible.[114] Scientists proffering magic tools to "solve the world's energy problems," including techniques to turn carbon dioxide into liquid fuel, have had to retract their results after reproducibility studies failed.[115] Recent surveys across hundreds of scientists found that over two-thirds reported they were unable to reproduce other studies' findings at least once.[116]

Around 85 percent of global health research spending is "wasted," directed to studies with opaque methods, incomplete reporting, avoidable design flaws and inaccessible publication.[117] The reason for this excess of dubious studies is not duplicitous researchers, but structures which encourage and incentivize poor-quality science. Academic careers, reputations and opportunities rely on prolific publication. But this "publish or perish" reward system, designed to improve research productivity and performance, has instead incentivized volume over validity.[118]

Such a context not only debilitates our understanding of the world but is increasingly and cynically manipulated by political forces wanting to disparage climate science.[119]

Thinking ahead

Given such vast obstacles, we need to develop a different vision of knowledge that can adequately navigate the magnitude of our systemic crises. These crises present questions, as physicist Alvin Weinberg noted, "that can be asked of science and yet which cannot be answered by science."[120]

A "normal" scientific problem may involve solving puzzles within a set framework.[121] But we face not puzzles, but problems with implications that are uncertain, time horizons that are unclear and responses that are fundamentally subject to value judgements. The binary of "true-false" is irrelevant when values and empirical facts can be legitimately joined in different ways.[122] Clarity is not so much a firm destination, but a negotiated dance of subjectivities and objectivities.[123]

This complexity means we cannot simply apply the narrow prism of technical expertise or the scientific method. Rather, we require new ways of thinking through problems that acknowledge the existence of diverse legitimate perspectives, rooted in different ways of seeing the world. Beyond reforms in academia, or campaigns to educate people on climate change, we need to start promoting what some scientists have called a model of "extended participation," which brings together scientists from different disciplines and participants from diverse communities of stakeholders.[124]

Rather than a science from above, pioneered by solitary individuals in removed universities, this vision sees strength in citizen collaboration and the democratization of inquiry.[125]

By acknowledging the importance of recognizing values, we can begin to question a science of absolute reason, which aims to distance itself as much as possible from emotion, spirit and ethics. As philosopher Masanobu Fukuoka noted, nature "grasped by scientific knowledge [alone] is a nature which has been destroyed; it is a ghost possessing a skeleton, but no soul."[126]

Instead, we should work toward a science with greater conscience, a dignified science. As the late Argentine embryologist Andrés Carrasco said, "we have to ask science for who and for what."[127] We need to pursue knowledge not for the sake of profit, but for the sake of understanding, and understanding for the sake of equity. Knowledge has to be wedded to responsibility.[128]

Pursuing this approach means acknowledging science as a plural and inclusive discipline, expressed in many different ways.[129] Our traditional conception of science conforms to highly narrow understandings of what constitutes scientific, and thereby environmental, knowledge.[130] We privilege the written word, and struggle to assign credibility to the oral, to the non-published, to that which lacks the accredited prestige of academia.

Ethnoecologists Victor Toledo and Narciso Barrera-Bassols confess:

> [As] scientists trained in the academic enclosures of modern science... rarely were we taught to recognize the existence of an experience, of a certain wisdom, in the minds of the millions of men and women who day by day work with nature... it is precisely because of that omission and forgetting on behalf of scientific investigation... that industrial civilization has failed in its attempts for realizing a more adequate management of nature.[131]

Scientific history is thus not just the story of brilliant scholarly discoveries, but also the rich history of local, peasant, indigenous and personal knowledges. But such expressions are rarely considered as knowledge, and are instead disqualified and dismissed as beliefs, faiths, opinions and superstitions.[132]

The canon of science is not a spontaneous archive of wisdom, but rather the historical set of ideas and thinkers that has been assembled, often through the exclusion of those deemed worthy of neglect: women, people of color and rural communities. The voices of the marginalized have been obscured. As Rodolfo Stevenhagen noted, through colonialism "the discovery of the Other" became the "concealing of the Other."[133]

The exclusion of these perspectives not only impoverishes our humility, but it fundamentally confines our perspective. Ecology demands an eye for dynamic and complex systems. Issues such as conservation, landscape protection, sustainable agriculture and water management are intricate, requiring many perspectives and routes of inquiry. Diverse lenses broaden the scope of our understanding. They allow us to ask richer questions, with richer answers.[134]

Our climate crisis hands us a precious opportunity to reformulate our epistemology. We need to break the separation between science and traditional knowledge, and to weave together a more plural and compassionate human understanding.

It is time to recognize the views and voices of the billions of human beings who every day interact with nature, treating them with the legitimacy, credibility and scrutiny applied to other epistemologies.

This does not mean essentializing or idealizing them. Romanticism can bear an uncomfortable likeness to racism; appreciation can quickly become appropriation.[135] The recovery of these histories is not an attempt to reify what seems exotic; rather it is the correction of the record through its completion. Peoples living close to nature have gained copious insights into agronomy, biology, botany, zoology, medicine, pharmacology and astronomy, among others.

Many communities have developed sophisticated taxonomies that surpass conventional methods. In the Amazonic region, for example, the Kayapó peoples use 16 terms to designate types of forest and savannahs.[136] The nearby Matsés peoples distinguish 104 types of primary forest and 74 types of secondary forest in their territory, a region the size of Bahrain.[137] The Machiguenga people, meanwhile, separate a total of 97 types of forest habitats, differentiating them by their soils, hydrology, seasonal regimes, water cycles, tree ages and flora.[138]

The extensive testimonies of local populations intimately connected with their environments predate established atmospheric science. In 1836, ethnographer Vadim Passek traveled across Ukraine, noting that "according to the observations of old-timers, the climate of Kharkov province...has become more severe, and it is now exposed to more droughts and forests."[139] Various Aboriginal Australian oral histories record sea level rises taking place over 7,000 years ago; cross-checking studies by marine geographers have found remarkable consistency between these testimonies and factual records.[140]

Inuit observations of Arctic changes, animal behavior and biological cycles have shown distinctive accuracy.[141] Generations of indigenous farmers throughout the Andes analysed the Pleiades star clusters to anticipate weather trends, observing the brightness of stars to predict rainfall. This historic method of meteorological prediction has been found to be as accurate as sophisticated computer modelling.[142] Indigenous folklore has often proved invaluable for academic historians, recalling stories omitted by written records.[143]

Whether it be the perception of animals as sentient and intelligent beings, the appreciation of the extraordinary capacities of plants or the interconnected nature of ecosystems, many of the most "novel" findings in the natural sciences have been long understood in diverse indigenous cosmologies.[144]

Valorizing these cosmologies, and redressing the processes that have subverted them, is also fundamental in the struggle against climate violence. From the carbon-absorbing benefits of traditional soils to the forest-enhancing capacities of local peoples, elevating these techniques and enshrining the rights of peoples to be able to practice them will be essential if we are to address the current crises.[145]

The possibilities of our future lie partly in our past, in the wealth of time-tested methods. We have much to learn about sustaining life from those with the greatest experience of grappling with this challenge. Our breaking civilization is only a few hundred years old; the Pygmy civilization of the Congolese forests is over 60,000 years old.[146] Highland communities in Papua New Guinea have endured for over 46,000 years.[147] As author Daniel Munduruku explains:

> Indigenous diversity is not a problem, it's a solution. For thousands of years [indigenous communities] have lived, collaborated, believed in, interacted with, and performed rituals in their places of origin. They know how to care for them, how to relate with them, how to gain sustenance without degrading.[148]

These are not bodies of abstract knowledge, but lived understandings integral to daily life and practice, that can teach us much about how to suture the divide between what we know, and what we do.[149]

The Moken people live among the Andaman Islands and are known for their nomadic life in the seas. In late December 2004, a major tsunami barrelled across the Indian Ocean, directly hitting the Andaman Islands. But unlike other parts of the islands, which suffered devastating losses of human life, only one member of the Moken community perished.

They owed their salvation to their inherited warning system. Elders had passed on fears of the *Laboon*, an enormous wave capable of devouring all in its path. Only those high in the mountains, or in deep waters, could survive it.[150] On 24 December, as tides receded and cicadas fell silent, elders noticed the signals. Moken communities fled into the hills. They were not the only ones. The Urok Lawai, the Ong and communities on the island of Simeulue all headed inland, attentive to the signals around them.[151]

▪   ▪   ▪   ▪   ▪

The story of climate change is one of the rise to dominance of a particular human relationship with nature, defined by callousness. Today we are blinded by that relationship, locked in its logic of devastation. The fixation of one expression of human nature teaches us that destruction is both our identity and our destiny. We assume that our malice is like air, inextricably part of the background.

But humanity as a whole has not generated this crisis. The imposition of a particular worldview has crushed our other options. A small portion of humanity destroys, but the entire human community pays the consequences. Just because we are in the same sinking boat, it doesn't mean that we all built it. Our extreme separation from nature is both recent and atypical. We can overcome it.

PRESENT

CHAPTER 7

THE GREAT BURNING

OUR CONCERN AROUND climate change centers around rapidly rising temperatures. Since 1750—the beginning of the industrial era—the average temperature on our planet has increased by around 1.2°C.[1]

That might seem like a small amount. But think of the planet's atmosphere—a vast stretch of space. Think of the planet's oceans—an immense body of water.

Humans have caused enough upheaval to heat the entire world—its seas and its skies—by over an entire degree Celsius. Just that one degree of temperature has been enough to trigger intense weather shifts, major polar melting and significant rises in sea levels. In India, for example, studies have shown that a rise in temperature by less than one degree Fahrenheit between 1960 and 2009 increased the chances of heat-related death by 146 percent.[2]

To illustrate the intensity of that upheaval, the astrobiologist Caleb Scharf offers us a useful thought experiment: let us visualize our emissions as a burning forest.[3] Between 1751 and 1987, our burning of fossil fuels dumped 737 billion tonnes of CO_2 into the atmosphere. Between 1987 and 2014, over 743 billion tonnes were dumped. So from 1751 to 2014, human beings pumped a total of 1,480 billion tonnes of CO_2 into the atmosphere.

When trees burn during a wildfire, every hectare of scorched forest releases about 1.6 tonnes of carbon dioxide into the atmosphere. In order to release an equivalent to the CO_2 emitted over the past 263 years, we would need around six million square kilometers of forest to burn every year during that time. In other words, an area of land only a little smaller than Australia, would have to burn every year for over two centuries.[4]

But let's draw out those numbers in a different way. Today we dump around 40 billion tonnes of carbon dioxide into the atmosphere every year. That's the equivalent of 42 million square kilometers of forest burning each year. To put that into context, the entire African continent makes up a little over 30 million square kilometers. Picture that: our current emissions, seen as a forest fire, equate to a territory of land greater than Africa burning each year.[5]

We can also illustrate our influence in other ways. Through the greenhouse effect, reinforced by human influence, our atmosphere and oceans are accumulating great amounts of additional heat every second. If we put that into numbers, our planet is holding back 250 trillion joules of extra energy every second. When an atomic bomb was dropped on the city of Hiroshima, it unleashed a wave of unfathomable destruction. In energy terms, it discharged 63 trillion joules of energy. Today, we are building up heat at a rate of four Hiroshima detonations a second.[6] Every 10 days, a billion tonnes of carbon dioxide are released into the atmosphere.[7]

Climate change is part of the Earth's history. Shifts in the planetary orbit, the cycles of the sun and volcanic eruptions are just some of the many influences that have shaken and shifted our climate system over time. But contemporary climate change is characterised by an unprecedented pace of change, resulting from human activity. Over the last centuries, a small portion of humanity has, knowingly and unknowingly, engineered a perilous outpouring of heat-trapping and climate-warming gases into the atmosphere. The ongoing imprint of those actions implies deep changes to the planetary and social balance.

Atmospheric basics

Life needs energy to be. Our body develops by ingesting food, plants grow by feeding on sunlight. Every second, the sun—our nearest star—showers the world with energy. Sunlight drives the water cycle, energizing the movement of the seas and evaporating water, which rises into the skies to return as rain. The tilt and turn of the planet distributes sunlight unevenly around the Earth. This thermal gradient, the temperature difference between tropics and poles, drives the circulation of air.

If we translate solar energy into numbers we can measure, the Earth receives 173,000 terawatts a year. To illustrate the scale of this energy, compare it to the total photosynthesis of plants, which captures 100 terawatts per year. Our global economy consumes around 14 terawatts of energy per year.[8]

The atmosphere of our planet is a thin envelope of gases. Some of those naturally occurring gases have heat-trapping properties, making them greenhouse gases. These are essential for securing life, for without them, the Earth's surface would be 30°C colder.

As the sun's energy beams over the Earth, around a third of it is reflected back into space. The rest is retained through the greenhouse effect and absorbed, mainly by our oceans. This vital process of reflection and absorption is what determines the temperature of the Earth.

But by radically remolding ecosystems, and emitting huge amounts of pollutants, we have changed the composition of the atmosphere; heat that would otherwise escape is accumulating. This has led to a significant difference between the solar energy absorbed by the Earth and that which returns to space.[9]

The equation is simple:

Heat in − heat out = changes in heat, changing temperature[10]

Let's imagine the atmosphere as a set of blankets wrapped around the Earth. By pumping more heat-trapping gases into the atmosphere, we have been adding blankets. The concentration (the amount, the thickness) of greenhouse gases in the atmosphere has been increasing rapidly over the last few hundred years. Since 1750, the amount of methane in the atmosphere has increased by 150 percent and the amount of nitrous oxide by 63 percent.[11] Each gas has a fingerprint, an isotopic mix which marks its relative origin; when studying the gases of our new atmosphere, we find increasing atmospheric levels of a type of carbon associated with combustion; that is, the burning of fossil fuels.

Knowing climate change

So how can we be certain about climate change? You might have heard this commonly cited idea: 97 percent of scientists agree that global warming is real and caused by humans.

The statistic is accurate, but its associated story is often misleading. Scientific understanding originates not in the opinions of individual scientists but rather in the accumulation of proof. The thesis that human beings are responsible for global warming and resulting climate change is based not on a consensus of experts, but on a consensus of evidence.[12]

The figure of 97 percent comes from a 2013 study. The study's authors, led by researchers John Cook and Dana Nuccitelli, examined 11,944 papers on the climate published from 1991 to 2011. Of those papers that stated a position on global warming, around 97 percent concluded that climate change is real and caused by humans.[13]

That is, thousands of diverse studies on animal migration, oceanography, climatology, geophysics, biogeography and agricultural science all observed noteworthy changes and pointed to one culprit: human beings. The remaining three percent of studies pointed in different and often contradictory directions. Detailed analyses of that minority of studies found recurrent methodological shortcomings and major issues with reproducibility.[14] Subsequent revisions of the 2013 study have shown that the number of scientific papers validating the thesis of anthropogenic climate change may be closer to 99.94 percent.[15]

Now what does that "consensus of evidence" look like in greater detail? Wherever we look, we find transformation. We see it when we look at sea levels, rain patterns, the spread of polar ice, the dates on which seasons arrive, patterns of animal migration, thermal surface emissions and vertical profiles of humidity.[16] There are tens of thousands of lines of evidence all leading in one direction.

These strands of evidence are drawn from a variety of sources—satellite data, independent temperature estimates, thermometers, ocean buoys, underwater temperature probes—which are marking very similar trends.[17] When we add these multiple datasets to centuries of recorded measurements and millennia of oral histories, and apply our knowledge of rudimentary physical and chemical cycles, a common conclusion emerges: (a portion of) human beings are responsible for current climate change.

The scientific method is not only about accumulating proof, but about disproving doubt. As the evidence has piled up, we have gradually discarded alternative theories that might also account for the extraordinary levels of global warming today, such as natural variation, solar radiation, volcanic activity and carbon cycle swings. Today we know that responsibility lies firmly at the feet of human beings. Without human intervention, it is likely that natural processes would have cooled temperatures in recent decades.[18]

Although the core of the theory is solid, there is much we are uncertain about. Climate science's impossibly ambitious endeavor—to understand the dynamics of the Earth system—means it will always fall short.

What we don't know: Uncertainty and humility

Anyone interested in the science and politics of climate change can quickly get lost among labyrinths of data, acronyms, graphs and statistics. To help us sift through the reams of information, and grasp the contours of our climate crisis, we need to be able to critically read and comprehend scientific research.

To understand how our climate works and will work in the future, scientists rely on a range of tools. Their primary instruments are numerical climate models. Climate models allow scientists to run hindcasts and forecasts, tests that map the past and predict the future. Historical runs are helpful to test current climate models and to determine what roles different factors have played in shaping weather patterns. These factors, from greenhouse-gas emissions to changes in the Earth's reflectivity, are known as radiative or climate forcings.[19] Hindcasts can also help us disentangle the causes of particular episodes of extreme weather through a process of attribution, while forecasts are indispensable to help outline the potential scenarios of the future.[20]

These models are composed of equations, which try to represent the processes that shape our global climate, including the movements of the atmosphere, the tides of ocean currents, seasonal cycles and carbon flows. The equations plot the biophysical principles that underlie our world, from laws of motion to laws of energy. Depending on their scope, they are fed with different inputs: empirical observations such as land topography, solar irradiance, the bathymetry of the ocean and the extent of vegetation.[21]

These equations are primarily partial differential equations, tremendously complex multivariable equations that cannot really be solved, but rather approximated numerically.[22] Scientists are forced to make empirical estimates, known as parametrizations, to realistically incorporate complex processes into models.[23]

Climate models are made up of tens of thousands of equations, comprising millions of lines of code that require supercomputers to run.[24] As historian of science Paul Edwards notes, "no fields other than nuclear-weapons research and high-energy physics have ever demanded so much calculating capacity." These models are constantly tested, calibrated, tuned and refined, as new insights emerge.

But although the models at our disposal are increasingly sophisticated, they are always limited and imperfect attempts at comprehending the infinite intricacies of the Earth system. We can understand this intuitively when we see how unpredictable daily weather is. Although forecasts can be made, the exactitudes of the future are incalculable. The climate is a chaotic system, where small shifts can ripple into large consequences. The concept of the butterfly effect, coined by meteorologist Edward Lorenz, asks us to imagine whether the flap of a butterfly's wings might affect the formation of a tornado.

Such ubiquitous complexity makes climate modeling very difficult. The easiest scientific models are linear, incarnating consistent, proportionate and predictable processes.[25] But all the key elements in our climate—air, water, ice—are nonlinear. They can shift rapidly and unpredictably, with transformative consequences for the models. This is the inherent difficulty of analyzing systems of life: they are constantly changing and evolving.[26]

Climate scientists bear the tremendous challenge of accounting for tricky phenomena, including the movement of seas, temperature fluctuations across millennia, solar radiation and the Coriolis force (Earth's rotation). Clouds in particular, continuously forming and morphing, evade easy simulation. It is very hard to accurately mathematize natural variability, complex behaviors of retained gases in the Earth system and atmospheric processes such as humidity.[27]

The sheer scale and nonlinear properties of the global climate prevents us from applying conventional scientific procedures. We cannot perform experiments to replicate the Earth, or easily isolate variables in a laboratory setting. The dimensions simply surpass our gaze.

Further uncertainties abound. Our geological and meteorological record is fragile and incomplete, given that comparable and consistent observations are only relatively recent. There are also observational uncertainties, as gases associated with farming, such as nitrous oxide, are, for example, tremendously hard to measure. Clearly distinguishing human sources from biogenic sources is similarly precarious. Some of the most intricate climatic processes, which are difficult to quantify or comprehend, have to be excluded from methodologies. Many climate models don't account for fast feedbacks (melting ice), or long-term carbon cycle feedbacks (permafrost methane release).[28]

Forecasting the trajectory of climate change is further complicated by the fact that human behavior is as nonlinear as the Earth system it unfolds in. Demographic changes, political decisions and future energy use—vital influences on our climate prospects—are all deeply uncertain. While climate models have been substantially effective in predicting climate impacts, they have been scarcely able to depict future impacts on ecosystems and societies.[29]

As a result of all these inherently unpredictable dynamics, climate models suffer from what is known as "irreducible imprecision."[30] All scientific assessments of our ecological past, present and future, as philosopher of science Jeroen Van der Sluijs notes, are "unavoidably based on a mixture of knowledge, assumptions, models, scenarios, extrapolations, and known and unknown unknowns."[31]

There is no sign of this imprecision disappearing any time soon. In 1990, the Intergovernmental Panel on Climate Change (IPCC) predicted that our uncertainties regarding the climate would diminish. But the opposite has in fact occurred. Greater knowledge has brought greater knowledge of our ignorance. Our ranges of uncertainty have expanded as more scenarios, processes and possibilities are integrated into calculations.[32]

When we add up all the sources of incertitude, the size of our aggregate uncertainty rises to the fore.[33] Our capacity to understand the climate, and its interaction with human societies, will always be catching up to systemic dynamics. This reality should strip us of any illusion of clairvoyance. We cannot realistically calculate economic, social or ecological consequences decades or even years away. When it comes down to it, we simply don't know how our planet, our home, is going to precisely respond to the emissions we have unleashed. Owing to this abundant complexity, climate science deals in ranges and probabilities, not in strict determinations.

Interpreting uncertainty

Acknowledging uncertainty should not lead us to embrace denialism, or nihilistically cast all models aside. We have ample insights into the prospects of the future. As climate modeler Gavin Schmidt clarifies, "[w]e may be just as unsure as before, but we are unsure on a much more solid footing."[34]

It is by appreciating the limits of research that we can begin to see its plentiful value. As the statistician George Box advised: "all models are wrong, but some are useful."

Uncertainty may feel uncomfortable, particularly in political contexts that demand the vigour of clarity. The limitations of atmospheric science have often been invoked to justify a delayed response to climate change. The argument goes that, since we don't know what will happen, there's little reason to expend resources and energies on addressing what may not occur, or may not be as bad as expected. Indeed, many corporations and governments have purposely worked to magnify, inflame and manipulate the uncertainty of climate change in their interests. But we should be drawing the opposite conclusion. Uncertainty should be a fuel for action not passivity.

Most of our previous studies are underestimates, having excluded sensitivities and nuances of which we are now aware. In recent years, we have revised and increased our predictions of sea level rise, hiked estimates for polar melting and learned that the Earth historically is more responsive to changes in CO_2 than we had previously thought.[35]

We have discovered that certain ice sheets, from Greenland to Antarctica, are far less resistant to warming than expected, and as a result, the likelihood of faster melt, and larger sea level rise, is higher.[36] We have clarified how tree trunks emit methane, understood better how hydroelectric dams build up greenhouse gases, realized that global livestock emissions are significantly higher than previously calculated and determined that mainstream scientific reports are undercounting the rate at which land use is changing.[37]

We have uncovered manufacturers which have intentionally undercounted their emissions and major loopholes in the official process by which countries account for their emissions.[38] The incorporation of missing data into models has also honed our picture of the past, showing even faster warming trends.[39]

Changes in the real world have outpaced our own expectations. The Arctic has lost sea ice decades ahead of its anticipated trajectory.[40] When the first IPCC report emerged, it did not mention the potential collapse of the Antarctic ice sheet, or the transformation of thermohaline circulation; such scenarios are, however, routinely considered today.[41]

And among all the various imperfect models that we have assembled, those that have proven to be most accurate in their simulations have been those that predict the highest levels of global warming.[42]

▪ ▪ ▪ ▪ ▪

Despite such concerning findings, in this context of irreducible uncertainty, we have tended toward restraint. The majority of atmospheric science is characterized by reticence, understating risks and providing the most hopeful projections. Analyses of past climate studies have found that they are largely "conservative in their projections of the impacts of climate change," erring "on the side of least drama."[43] Scientific norms of "restraint, objectivity, scepticism, rationality, dispassion," although useful in many contexts, have been signaled as culprits guilty of diluting the urgency. Scant attention is paid to "fat-tail risks": high-risk possibilities, with lower probabilities.[44]

But as atmospheric scientist Matthew Huber outlines: "when it comes to evaluating the risk of carbon emissions, such worst-case scenarios need to be taken into account. It's the difference between a game of roulette and playing Russian roulette with a pistol. Sometimes the stakes are too high, even if there is only a small chance of losing."[45]

CHAPTER 8

UNDERSTANDING EMISSIONS: WHERE, WHO, WHAT, WHEN AND HOW

BEYOND THE RICH DEBATES surrounding complexity in climate science, when it comes to emissions, there are five major discussions: where, who, what, when and how? Where are emissions coming from? Who will make the emissions cuts? What level of emission cuts do we make? By when should those emission reductions take place? How will they be made?

Where: Types of emissions

Our excessive emissions originate in five primary sectors: energy, food, forests, transport and industry. These five sectors can be further condensed into two key sources: where we get our energy from and what we use our land for.

Energy: Transport, electricity and cement

Responsible for two-thirds of emissions, energy is the protagonist of climate change. The global economy is around 80 percent dependent on power from fossil fuels (coal, oil and gas). The remaining power is largely made up of nuclear, hydroelectric and traditional biomass energy. Solar and wind energy account for a mere six percent of global electricity generation, and less than two percent of energy.

Our fossil-reliant economy burns over 10 billion tonnes of fossil fuels every year. When fossil fuels are burnt, whether in car engines or power stations, they emit large amounts of greenhouse gases, particularly carbon dioxide. Out of all fossil fuels, coal has the highest carbon intensity, meaning that if you burn a tonne of coal (versus a tonne of oil), you'll get a lot more carbon dioxide coming from coal. In theory, gas has a lower carbon footprint, but due to rising demand, natural gas is today responsible for more carbon dioxide emissions than coal.[1]

Fossil fuels also cause emissions through leakage. As gas is transported through tubes and pipes, it seeps out into the atmosphere. Nearly a third of Russia's emissions come from leaking pipelines that emit methane.[2] Abandoned oil wells are also significant sources of methane. Methane, when compared to carbon dioxide, has what scientists call a higher global warming potential: a measurement showing the ability of a gas to trap heat.[3] Over a 100-year time period, methane has a warming potential 34 times more potent than carbon dioxide.[4]

With such potent pollutants seeping daily into the atmosphere, we need to rapidly end our dependence on fossil fuels, transforming our transportation, industrial and electric systems.

Although it is related to energy, cement production is generally accounted for as a distinct source. Cement is the core ingredient in concrete, the building block of many of our homes, offices, hospitals and bridges. Cement manufacturing is currently responsible for around five percent of global emissions. The most common form of cement is made of limestone and aluminosilicate clay, which are mixed and baked in kilns. When exposed to heat, limestone's calcium carbonate converts into calcium oxide and carbon dioxide, in a process called calcination.

Land use: Forests and food

Greenhouse gases are not only emitted but are also held in sinks, where they are processed or stored. These are essentially the concentration spots of the world's stored carbon. There are many types of natural sinks. Oceans are the largest stores of carbon. Next comes soil, the deposit of three times more carbon than all the flora (trees and plants) on the planet. Forests, which inhale carbon dioxide, are also vital sinks.

But many sinks are destroyed or disrupted through land use changes, such as forest clearance, urbanization, mining or road construction. When peatlands are converted to fields, for example, dried peat emits carbon. When forests are logged and razed to make way for plantations, their previous carbon-absorbent capacities are extinguished.

Land uses also carry emissions of their own. Today, farmed fields make up two-fifths of the world's land surface, generating emissions through biomass burning, the fertilization of soils (nitrous oxide), the flatulence of livestock (methane), the emissions of farming machinery (carbon dioxide) and rice cultivation (methane), among others.

When all the diverse impacts on the land are added up, land use emerges as the second most significant contributor to climate change, accounting for around a third of global greenhouse-gas emissions.

Who: Emissions and authorship

The popular mythology of climate change holds that humanity as a whole is responsible, that all human beings are to blame. But we are not all equally responsible. Climate change is inseparable from global inequality. The history of carbon is one of unequal power obtained through unequal pollution.

Who is responsible? If we assign responsibility by country, the United States carries 40 percent of world emissions debt. There are also "carbon creditors": states whose share of CO_2 emissions has been smaller than their share of global population, including Bangladesh, China, India, Indonesia and Pakistan.[5]

In 1825, Britain was responsible for 80 percent of all emissions from fossil fuels; by 1850, it was still responsible for 62 percent.[6] The "richest states," despite having less than a fifth of global population, have been responsible for four-fifths of historical carbon emissions. Until 2000, the United States had emitted 27.6 percent of historical emissions, while Nigeria had emitted 0.2 percent and Brazil 0.9 percent.[7] Today, El Salvador's average emissions per capita are 45 times lower than the Qatari average, and 15 times lower than the US average.[8]

If we blame the companies where the emissions originated, then the figures look somewhat different: only 90 companies are responsible for nearly two-thirds of all emissions since 1750.[9] Half of those emissions were emitted after 1988, by which time the threat of climate change was widely known.[10] One study found Exxon Mobil alone is responsible for 3.22 percent of global emissions between 1751 and 2010.[11]

But what if we allocated the blame to individual consumers? Oxfam reports that the richest 10 percent of the world population is responsible for half of global emissions.[12] The poorer half of the world is responsible for a mere tenth. History shows that the richest one percent have emitted around 175 times more than the poorest 10 percent.[13] The richest one percent of Saudi Arabians have an emissions footprint 2,000 times greater than the poorest Malians.[14]

Neither is every molecule of gas emitted into the atmosphere identical; there is a difference between superfluous and subsistence emissions.[15] Some scholars have argued that a sixth of the global population should be exempt from responsibility, and excluded from emissions counts, given their minimal level of resource use.[16]

From every angle, a common fact emerges: not everyone has emitted equally. There are clear asymmetries, which spell out two general rules of climate inequality:

- The world's richest peoples tend to use more energy, drive larger cars, heat larger homes, take more flights and buy more things. Around 80 percent of the planet's resources are consumed by a fifth of its population.
- Inequality can exacerbate emissions. The more economically unequal a country, the higher its carbon pollution tends to be.[17]

And as we shall see later on, these rules run parallel to the laws of unequal impacts:

- Those most responsible for climate change are likely to be the most unaffected, and the most able to adapt.
- Those least responsible for climate change are likely to pay the highest costs and experience the strongest impacts.

Why does it matter who has despoiled the global commons of our atmosphere? One explanation is that our failure to find a way of adequately allocating responsibility is a prime reason for inaction. At every single environmental summit, the main disputes arise between the poorest and richest states over who shall shoulder what responsibility for reducing emissions and rectifying the consequences of climate violence.

This fight is not merely a technical disagreement, but a manifestation of a much deeper historical struggle. As Asad Rehman, a veteran civil-society representative at the climate negotiations, explains: "If you think the international climate negotiations are about the climate, you don't know what they're about. They're fundamentally about political economy."[18]

Since power has historically been obtained partly through carbon pollution, it serves the interests of the powerful, from states to companies, to deflect attention from the origin of their wealth. From the genesis of international climate negotiations, the richest delegations have worked assiduously to water down commitments and dilute the language of agreements. The IPCC's first assessment report was written with the help of 11 scientists from the oil and gas industry.[19] To this day, policy summaries of international climate-science reports are vulnerable to immense political pressure from polluters.[20]

Powerful states with big carbon debts have used all kinds of tactics to hustle other countries, scuttle accords, insert loopholes and sideline principles of equity. Ultimately, the United Nations is a reflection as an unequal world. The size of a country's diplomatic corps, the weight of its national economy and the strength of its lobbying power all define a state's ability to shape outcomes and evade consequences.

In the shadow of such inequality, summit after summit has yielded fitful progress. On some counts, we have regressed. The current architecture of the Paris Agreement, adopted in late 2015, relies on a system of "Intended Nationally Determined Contributions." In human terms, these are voluntary pledges to cut emissions made by individual governments, with little basis in science or justice.

The agreement is not binding, without firm obligations or penalties. In other words, it is little more than a collection of well-intentioned promises on paper. In 1991, during early negotiations around the initial UN Convention, a Pledge and Review agreement was proposed by Japan. It was swiftly dismissed, given a broadly held consensus that any agreement on climate change must firmly incorporate the historical responsibility of developed countries and contain legally binding obligations. As the former Indian negotiator Chandrashekhar Dasgupta would later observe: "[A]n approach that was summarily rejected as inadequate at the outset of the climate change negotiations, is being hailed today as a great advance!"[21]

The 1987 Montreal Protocol on ozone depletion, in contrast, was binding, committing the world to a complete phasing out of ozone-degrading substances. The Protocol has been dramatically successful in reducing the emissions of both chlorofluorocarbons (CFCs) and hydrochlorofluorocarbons (HCFCs).

The international climate regime has also effectively eroded principles of international fairness. When the United Nations first started meeting around climate change in 1992, it acknowledged an important principle: developed countries should do more to mitigate climate change, because they were historically more responsible, and were currently more technologically and economically capable. But following historic lobbying from the richest states, UN negotiations have gradually stripped away these commitments to justice.

What: Temperatures and targets

What is the limit of the global warming we can handle? The honest answer is there is no such limit. Already, the implications of our current levels of warming are devastating communities. Every step further is a step too far.

But politicians have agreed over the last years on two targets: staying below 1.5°C or 2°C of warming. These represent thresholds of danger that have been contested for years. Although these round numbers were politically agreed, they were broadly based on key scientific studies that highlighted the risks of temperatures beyond those thresholds. One IPCC assessment considered that temperatures above 1.9°C start running the risk of "triggering the irreversible loss of the Greenland Ice Sheet, which could trigger sea level rises around seven meters."[22]

Ultimately, each degree of warming, each decimal point, directly translates into a different degree of destruction. The higher the warming we are willing to allow, the higher the price of life we are willing to pay. Every deadline, every demand we make, every target we forsake involves a particular calculation of loss. The longer we wait, the less action we take, and the more we are willing to sacrifice. This is the fateful arithmetic of climate change: every number is in fact a code of life.

The question is not what level of climate change is dangerous, but rather, what level of warming will we accept? The debate around temperature is one around the acceptable level of devastation, the allowable volume of human pain. When we say that there is permissible warming (up to 2°C), we may be tacitly justifying all the episodes of climate change-fuelled weather we are seeing around the world.

Scientists have warned that 2°C "could cause major dislocations for civilization."[23] The difference between 1.5°C and 2°C is one of life and death: major gradients of tidal flooding, heat extremes, water shortages and falling crop yields.[24] The margins of these gradients are everything. For island nations, it is the difference between being submerged or surviving as a state.[25] For farmers, it is the difference between the semblance of a harvest or immiserating destitution.

The 1.5°C guardrail is, for example, the sole hope for preserving coral reefs, many of which need warming lower than 1.2°C to survive. Surpassing 1.5°C would destine large percentages of the Earth's surface to desertification.[26]

But already 1.5°C is largely considered unachievable without an immediate wholesale transformation of the global economy. Even 2°C, a target previously deemed catastrophic, seems out of reach. Currently we are on track for at least 3.2°C by 2100, with other studies warning of even higher likely temperatures.[27]

When: Too little, too late

Climate change is often framed as a ticking clock, a countdown. But we continue to push back our parameters. In 2007, Rajendra Pachauri, the chair of the IPCC, declared, "If there's no action before 2012, that's too late. What we do in the next two or three years will determine our future. This is the defining moment." After 2012 then passed, leaders again warned of a final moment. The reality is that there is no deadline. Every ultimatum is ultimate. There will always be warming to prevent, there will always be devastation to diminish.

Nonetheless, a common fear that surrounds climate change is that we have left it too late. At some point, we will reach the point of no return, entering a stage where the changes we have triggered in the atmosphere are no longer reversible. Our immense influence over the climate system will have transformed it into a system beyond our control. This idea is legitimate, rooted in a concrete idea of what is known as feedbacks.

The Earth is an interconnected system. From our soils, to our oceans, to our winds, everything is intertwined in the web of life.

Feedbacks are the mutual interactions between the elements in our climate system. Such intimate connections make every ecosystem relevant: what occurs in the Arctic, Amazon and Antarctic is closely tied to the fates of the world's peoples.

There are broadly two types of feedbacks: positive and negative. Positive feedbacks amplify an effect, while negative feedbacks diminish it. Those concerned about climate change are most worried about positive feedbacks that accelerate vicious cycles of greenhouse-gas release and warming.

Perhaps the most famous positive feedback involves albedo. Albedo represents the Earth's reflectivity—it is a measurement of a surface's ability to reflect light. The properties of albedo mean that dark seas absorb far more sunlight than white icecaps, which reflect it. As ice melts, the albedo of the Earth is reduced, meaning less light is reflected and more heat is absorbed.

There are many other positive feedbacks, particularly involving the ocean. When we think about climate change, we usually think about how hot it is, or about surface air temperature. But only a strand of the heat retained by the greenhouse effect warms the air; 90 percent of the additional heat ends up in the oceans. The top few meters of the world's oceans can store as much heat as our entire atmosphere. Without the immense storage power of our seas, surface temperatures would be 36°C hotter.[28]

The ocean's uptake of heat has doubled in the past two decades.[29] Water expands when it heats, in a process known as thermal expansion. By absorbing so much heat, the temperature of the oceans rises. Rising ocean temperatures increase the risk of polar melting, which increases the amount of water in the ocean, which increases sea levels, that in turn lift ice formations, inviting in warmer water, causing further melting, and further sea level rises.[30]

In the atmosphere, warmer air evaporates more water, leading to the greater presence of water vapor, the most prominent greenhouse gas. Hotter temperatures mean more forest fires, which burn more trees, emitting large amounts of CO_2 and reducing the number of trees absorbing CO_2. In some contexts, intense heat leads to more rain—which leads to more plant growth—which leads to more kindle—which leads to stronger wildfires—which leads to further heat.

Crucial feedbacks are found in relation to greenhouse-gas sinks. From Scandinavian mires, to Congolese peatlands, to clathrate hydrates stored under ocean sediments, as temperatures warm, many ecosystems face unprecedented carbon and methane release.[31] The permafrost in particular contains 1.5 trillion tonnes of frozen carbon, twice as much as the atmosphere. Yet through rising temperatures, the permafrost is being defrosted, releasing methane gas from underground, which is in turn exacerbating global warming, which accelerates permafrost melting. Current trends suggest we could be on track for losing all the top three meters of permafrost across the world by the end of the century.[32]

Human communities also shape feedbacks. Countries and communities that suffer failed harvests from droughts may try to make up for the loss by expanding the amount of cultivated land. But expanding that land might require cutting down trees, further exacerbating the problem. As global temperatures rise, demand for air conditioning follows, increasing electricity demand which requires greater fossil-fuel combustion. As permafrost and icecaps melt, newly accessible lands can be opened to fossil-fuel extraction.

There are also negative feedbacks that dampen dynamics of warming. These ironically include the emission of aerosols, small particles that are thrown into the atmosphere during fossil-fuel burning. Another negative feedback is vegetation growth, caused by higher concentrations of carbon dioxide in the atmosphere. The surfeit of carbon dioxide has contributed to global greening, which has slowed the rise of CO_2 in the atmosphere.[33] Increased emissions and warming could be accompanied by plant growth in areas such as the Sahel and the Arctic.[34] But these negative feedbacks only exert a minor influence, and are vastly outweighed by all the amplifying feedbacks.

Nonlinear positive feedbacks are gaining influence in the Earth system, and can only be reduced by lowering the Earth's temperature. But, despite their importance, and the dizzying unpredictability they inject into the climate conversation, we continue to act as if climate change were a linear problem. Policymakers often assume we can over-emit for a few decades but then recover lost ground through ambitious future reductions.[35] This idea, known as overshoot, has been accepted as normal. It allows for the wholesale redefinition of climate targets and provides a tacit argument for procrastination.

The serious consideration of "overshoot" measures reflect our dreams of mastery, conceiving the Earth system as a static machine of inputs and outputs, easily fixable.[36] But the world is not a machine we can bend to our will. Nature is not a mechanical or "balanced" system. It is raucously unpredictable, defined by revolutionary changes and imbalances.[37] We cannot rebuild a collapsed ice sheet or reverse-engineer a feedback.

Instead, the colossal climate system is characterized by inertia: it takes time to adjust, for impacts to ripple through. Warming is gradual: first the air warms, then the land, then the surfaces of the ocean and then, slowly, the depths of the sea.[38]

Any complex system like the planetary climate is defined by "ubiquitous delays," where actions can often take time to translate into observable impacts.[39] The impact of pollution lags behind its emission. Climate change has its own momentum. Emissions are cumulative, taking time to translate into change. The Earth system adjusts to shifts over long time scales. The real scale of sea level rise, acidification and atmospheric temperature produced by current emissions will reveal themselves only gradually.[40]

The spate of climate extremes emerging today is a response to emissions from decades ago. Our present is the effect of atmospheric memory, a visible wound of historic trauma. We are seeing the warming of the 19th and 20th centuries unfold before our eyes.

The emissions we have already pushed into the atmosphere, the scars we have carved into the Earth, have already set into motion unavoidable changes. The emissions of today are legacies to be written into the future, burdens dropped on future generations.

It will take centuries for the full reaction to current emissions to be expressed.

The longevity of carbon's memory makes it particularly insidious. What is emitted can only be drawn down over long time scales. Even if we were to freeze all emissions tomorrow, the impacts of unleashed climate change would extend hundreds of years into the future. Most of the gases in our atmosphere will be there in hundreds of years.

The inertia of the carbon cycle is coupled with human inertia: our inability as societies to respond rapidly to problems, burdened by the baggage of the past. The IPCC has previously warned of "the tendency for past decisions and events to self-reinforce, thereby diminishing and possibly excluding the prospects for alternatives to emerge."[41]

Every investment made is a veiled stake in the future. Every factory opened, every pipeline laid down, every technology entrenched cements a particular reality. Existing infrastructure commits the world to significant warming.[42]

Warming and deep changes to ecosystems are locked in.[43] Some western Antarctic ice sheet glaciers have gone beyond the point of return, facing unstoppable collapse.[44] Even if emissions were to cease today, there is a chance that current levels of emissions commit us to reaching 1.5°C.[45]

How: The carbon budget and the roadmap

Given the inertias and feedbacks of the systems, the main imperative is to avoid a situation where thresholds (tipping points) are breached. At this point, we will have lost the ability to stabilize the climate. Feedbacks loops, mechanisms beyond our control, will become protagonists of the Earth system.[46] An era of endless warming will have been set into motion. We will have ignited the engines of extinction.

Now, if we want to prevent this situation, what do we have to do? Given that the concrete problem is excessive emissions, the intuitive solution is emissions reduction. This is half-right, but the problem is that huge emissions over decades have increased the concentration of gases in the atmosphere. It is this concentration, this accumulation of gases, that needs to decrease. Even if all emissions ended tomorrow, we would need time for carbon sinks to process and absorb carbon dioxide.[47]

In any case, the amount of greenhouse gases pumped into the atmosphere must be rapidly reduced to zero, and then, beyond that, become negative. Negative emissions occur when the amount of greenhouse gases absorbed by the Earth exceeds what is emitted.

Future emissions need to be very limited. Every single tonne of carbon dioxide or methane that we emit translates into a particular amount of warming in the atmosphere. Every single tonne of gas we preclude means an avoided burst of warming.

If we want to stay below the guardrail of 1.5°C, we only have a limited amount of carbon dioxide left to emit before we transgress our boundaries. That amount is typically called a carbon budget.

Our global economy emits around 40 gigatonnes of carbon dioxide every year. To stay below 2°C, we have a carbon budget left amounting to somewhere between 150 and 1,050 gigatonnes.[48] These budgets cannot be exceeded if we are to avoid breaking the temperature goal.

How do we keep our carbon budgets in check? We need tremendous reductions of emissions, achievable through modifying our food system, shifting our energy sources, reconfiguring our modes of transportation and transforming our economy. We will also have to preserve, restore and expand the sinks of emissions, by increasing the carbon retention of soil, reversing deforestation and curtailing desertification. Emissions ultimately have to go negative: carbon has to be absorbed by ecosystems at greater rates than it is emitted.

Yet we are nowhere near this level of action. Instead, we are seeing some of the highest rates of carbon-dioxide growth on record.[49] Many of the world's richest states, who bear the largest responsibilities for climate action, are postponing previous targets and policies.

Literacy and ambition

If we are interested in ambition, we need to be able to detect it. Ecological literacy requires the ability to examine any climate proposal or action plan with a critical eye. Ultimately, we should always recall that every study will be a servant to its background assumptions and starting points. Models, as mathematician Cathy O'Neil writes, "are opinions embedded in mathematics."

Take a relatively simple issue, such as how we define the starting point of global warming. The Paris Agreement, as the majority of the most prominent climate studies, uses temperature baselines of 1875, referred to as "pre-industrial conditions." Yet the Industrial Revolution began earlier, in the mid-18th century. Keeping the baseline at 1875, instead of 1750, excludes over 100 years of global warming. If we adjust models to incorporate the warming of those dozen decades, we have a much smaller greenhouse-gas budget, with 40 percent less carbon, and even less time than expected to tackle the problem.[50]

Another concept of particular relevance to climate models is equilibrium climate sensitivity (ECS). Climate sensitivity is a function which estimates how the Earth's climate system responds to a doubling of carbon dioxide in the atmosphere. ECS is used to assess the reactions of the climate and, by proxy, to determine how severe climate change will be. These readings are then used to inform temperature targets and build carbon budgets. Depending on the ECS value we use, we can get very different results. If we extrapolate from the climate commitments of the Paris Agreement, the world may face warming between 3°C and 5°C, in accordance with diverse ECS readings.[51] Recent research has suggested, however, that climate sensitivity could be far greater than the medians used for most modeling, potentially driving global warming past 7°C this century.[52]

Similarly, whenever we see any statistics around emissions reductions, we need to be attentive to their origin. Since 2007, China has been the country with the highest level of emissions, having overtaken the United States. But what does it mean to say that particular emissions are Chinese? China's rise to the top of emissions rankings coincides with its emergence as the world's workshop. As the richest countries deindustrialized, China industrialized, becoming the world's largest consumer and producer of energy. Nearly half of China's total emissions in the period 2002–2008 were generated in the export sphere.[53] But although those goods are consumed elsewhere, emissions involved in their production are registered as Chinese. As Dale Jiajun Wen observed, "[in] essence, China as the kitchen, while the west is the dining-room."[54]

Over recent years, many governments have proclaimed they have decoupled their emissions from growth. But studies show this "decoupling" is misleading, for when imported materials are accounted for, rich nations have not reduced material consumption.[55] Instead of progress on emissions reductions, in many of the world's richest countries, we have seen processes of carbon offshoring, where emissions are displaced through the importation of carbon-intensive goods.[56]

These may seem like unnecessary technicalities, but ultimately, the functions we use, the definitions we delineate, the timelines we stake and the assumptions we make do much to outline the actions we consider acceptable or effective. Across many models, lavish assumptions and devious accounting tricks are inserted to allow for states to procrastinate on real action, and pursue slower trajectories, at the expense of those most vulnerable to climate violence.

We have left things way too late to continue adjusting our ecological knowledge to the status quo, instead of adjusting our status quo to our ecological knowledge. We have simply run out of time for further excuses and vacillations. Consider the conclusions of these recent studies:

- We have only three years left before we fully deplete the carbon budget for 1.5°C.[57]
- Delivering the goals of the Paris Agreement implies ending new fossil-fuel extraction. The amount of carbon locked within existing fossil-fuel development projects exceeds the amount of burnable carbon permissible under IPCC carbon budgets.[58]
- To meet the Paris Agreement's goals, fossil-fuel emissions need to be halved every decade, reaching zero by 2050. Net emissions from land use will have to dive to zero. Carbon-dioxide removal technologies will need to scale up massively to the extent that they will have to pull 5 gigatonnes of CO_2 from the atmosphere (double what soil and trees do already).[59]
- Following international principles of equity, to stay under 2°C, rich nations will have to reduce emissions by 8–10 percent every year. Such reductions are only realistically feasible through the reduction of economic activity: reversing economic growth and downscaling the economy.[60]
- Staying below 1.5°C means phasing out all fossil fuels by 2030.[61]
- To keep temperatures below 2°C, no new gas, oil or coal-fired power plants can be built beyond 2017. At the time of writing, around 1,500 coal plants are currently under construction or under planning around the world.[62]

With such gargantuan tasks ahead of us, the probability of safety certainly feels extremely small.[63] It is likely that staying below 2°C, let alone 1.5°C, is only possible in models.[64] There is simply no patient or gradual way of reaching these targets. One scientific paper outlined the impossibility of preventing a temperature rise of 2°C "within orthodox political and economic constraints."[65] We essentially need to enact deep emissions cuts, eradicate deforestation, expand low-carbon technologies and restore ecosystems to remove carbon dioxide from the atmosphere. Failure to comprehensively accomplish any of the above makes any of the guardrails virtually unreachable.

With every day that passes, emissions continue, and windows of opportunity continue to close. Every delay is disastrous. The only clear answer is an acceleration of action. In a world of accumulating emissions and runaway impacts, speed is synonymous with survival.[66]

Radical change is therefore inevitable, either by the atmosphere or by humans who depend on it. But one thing is clear: accepting the depth of our ecological crisis involves acknowledging the inadequacy of our dominant economic model.

CHAPTER 9

THE POVERTY OF WEALTH: ECONOMICS AND ECOLOGY

We are in the 21st century, trying to resolve problems
of the 21st century, using economic theories and paradigms
from the 19th century. Today, we no longer have 19th-century physics,
19th-century anthropology, or 19th-century medicine.
But in economics and development we have stayed there.
MANFRED MAX-NEEF

* * * * * * * * * * *

There are two breads. You eat two. I eat none.
Average consumption: one bread per person.
NICANOR PARRA

* * * * * * * * * * *

The Project that threatens life does not respect boundaries,
that's why they call it Globalization.... Not only are our cultures,
our communities, our peoples and families at risk. It's worse—
life itself runs the risk of being destroyed.
Public Proclamation of the
Indigenous and Popular Indigenous Congress, Colombia, 2005[1]

* * * * * * * * * * *

OURS IS A WORLD of abundant inequities. The majority of our fellow human beings are deprived of one or more of our basic needs: clean water, nutritious food, affordable housing, sanitation, energy or adequate healthcare. As many as 4.2 billion people, around three-fifths of the global population, live in poverty, a rate which has increased dramatically in recent decades.[2]

Across the board, our current economic models have broken their generous promises: to end extreme poverty, to drive hunger into history, to deliver extensive well-being and lift all boats through their rising tides.

The story of pervasive progress is slowly crumbling. Advances in areas such as public-health promotion or the eradication of hunger have diminished or even been reversed in recent years.[3] In the United States, life expectancy has declined for the first time in decades.[4]

Over 50 nations are poorer than they were 30 years ago.[5] Between 2009 and 2013, the number of Europeans living with "severe material deprivation" rose from 7 million to 50 million.[6] In Italy, the number of people living in extreme poverty has tripled over the last decade.[7]

Income inequality has risen in virtually every country in recent decades.[8] Today, five men control the same amount of financial wealth as the remaining half of the world's population.[9] The richest fifth of humanity receives 70 percent of global income, while the poorest fifth receives 2 percent.[10] One American family, the Waltons, owns more wealth than the poorest 42 percent of the US population. In India, the top one percent of the population acquired 73 percent of the nation's new wealth.[11] The net worth of one individual, Bill Gates, surpasses 30 years of Haiti's GDP.[12]

The economic gulf between states is also expanding, translating into direct disparities. The location of your birth will shape the likely length of your life. If you are born in Japan, then an average life of 84 years awaits you. If you're born in Chad, you can expect a life 34 years shorter.

These numbers are eloquent reminders of a world adjusted to absurdity. Perhaps the most painful truths are the most simple. More money flows to purchasing weaponry than addressing the greatest expressions of misery. The annual fortunes of the world's billionaires could end extreme poverty several times over.[13] In September

2008, the Food and Agriculture Organization of the United Nations prescribed the sum necessary to eradicate hunger: $30 billion. Just months later, the governments of the world bailed out the banks with $17,000 billion. What was handed to the banks could theoretically have bought the world 600 years without hunger.[14]

▪ ˙ ▪ ▪ ▪

"Economics" and "ecology" share the same linguistic root: eco, from the word oikos (Greek for home or household). While the prism of economics historically looked at the management of resources on a domiciliary level, ecology focuses on the broader relationships that sustain the broader human habitat.[15] These goals are firmly intertwined, but over time, they have been wedged apart. Today, our dominant vision sees an economy that gains its strength through the suppression of ecology, obtaining prosperity through the impoverishment of people and ecosystems. As philosopher Bolívar Echeverría noted, our economy depends on "suffocating life and the world of life."[16]

Metabolism

Everything that surrounds you has a memory. This page, your seat, the light, your clothes. Every turn of the tap, every press of a button, every step on a pavement, every purchase is an interaction with a dense network of relations, symbolized by pipes, cables and supply chains.

The buildings that surround us are composed of cement and glass, which in turn require sand, limestone and dolomite, among other minerals. Food processors mix grains, fruits, vegetables and synthetic chemicals to make the products on supermarket shelves. Electronics are assemblages of copper, coltan, tin, nickel and dozens of other elements. Whether close or far, we are immersed in complex relationships.

Nature is the basis of the economy. The economy draws on life, and transforms it.[17] In the process, as materials are consumed and energy is applied, waste is released.

Some thinkers use the metaphor of a metabolism. Every organism has a metabolism, processing materials and energy. Like organisms, socio-economic systems exchange materials with the environment. These systems rely on this exchange to sustain and develop their own functions.[18]

▪　▪　▪　▪　▪

Thermodynamics, the study of energy, holds a basic law: matter is neither created nor destroyed. It also teaches us about the law of entropy, which holds that energy is transformed from ordered to less-ordered forms. A piece of wood is, for example, a low-entropy matter. Burning that wood turns it into high-entropy matter, in the form of gases.[19]

The economy nourishes itself on the energy of nature. But when nature enters the furnaces of industry, it is transformed and degraded. Trees are pulped into paper. Mountains are dismembered into mined minerals. In closed systems, metabolic processes produce waste.

Given these dynamics, and our reliance on the quality and longevity of ecosystems, we must be careful. The economist Herman Daly outlines three core rules for sustainability:

1. Renewable resources (fish, soil and groundwater) must be used no faster than the rate at which they regenerate.
2. Nonrenewable resources (minerals and fossil fuels) must be used no faster than renewable substitutes for them can be put into place.
3. Pollution and wastes must be emitted no faster than natural systems can absorb them, recycle them or render them harmless.[20]

When these rules are broken, ecosystems sink into imbalance and degradation. Climate change, for example, is an illustration of a violation of the third rule. Our climate crisis is, in very raw terms, a waste crisis. By weight, carbon dioxide is the greatest waste product of industries.

Daly's rules show us two major things. First, relationships matter. Our interactions with ecosystems shape those ecosystems, and the interactions we may have with them in the future. Questions of scale and resource use become important.

Second, the living world has limits. Our atmosphere, our soils and our seas are not invariably abundant sites of extraction, or limitless deposits for garbage. The rhythms and relations of nature place constraints on the possibilities of human societies. Given such parameters, the common concerns of economics and ecology turn on how we can thrive within environments without exploiting them. How do we use and distribute resources to ensure well-being and sustainability?

▪　▪　▪　▪　▪

For too long, the material limits of our world were deemed irrelevant among the priorities of economics. Nature was an object to be controlled, and humans were separate from any consequences of the project of dominion. The logics of growth, profit and wealth expansion trumped the stewardship of life. This neglect of boundaries, born of both blindness and calculation, has wrought a legacy of drastic destruction, embodied by the sheer strains borne by ecosystems. In our economy, renewable resources are uncompromisingly depleted, nonrenewable resources are plundered, and pollution is poured out relentlessly, smothering environments that struggle to absorb it.

We are overusing, expending and contaminating more than ever before. Our collective overconsumption, performed particularly by the richest groups, significantly surpasses the biocapacity of the planet. Our demands on the Earth's systems have doubled in the past four decades.

Such patterns show little signs of relenting, as the global metabolism continues its expansion. In the first millennium of the Common Era, the world economy grew at an average annual rate of 0.01 percent. From 1000 to 1500, it grew at 0.1 percent a year. Between 1500 and 1800, it grew at 0.3 percent a year. Between 1820 and 2000, that rate rose to 2.2 percent a year. But over the past two centuries, the global economy has multiplied by more than 50 times.[21] If the global economy were to grow by 3 percent a year until the end of our century, it would be 60 times larger than now in 2100.[22]

New kinds of economic thinking demand ecological realism: facing up to the planet's limits and underlying relations.

Prosperity

One of the major obstacles to ecological realism lies in the strength of orthodox visions of the economy, entirely segregated from issues of ecology. These framings offer us simplistic diagnoses and prognoses for the complex issues we all face.

Real issues of poverty and deprivation are framed as mere problems of scarcity: we are simply not producing enough. A mechanical link is established: success is found in material abundance and the pursuit of more. With more extraction, more production, more consumption, more services, more technology, more stuff, we will find more prosperity. This societal vision, known as productivism, is accompanied by consumerism: we can gainer better individual well-being through having more.

The promise of increased production and consumption underlies the ideology of much of our modern society. Businesses focus on swelling profits, appropriating greater market shares and expanding operations. Economists debate how to accelerate the engines of growth. Newspapers and broadcasters project the latest quarterly figures.

In politics, GDP is positioned as the standard barometer of well-being, the proxy of prosperity. Yet GDP is a tremendously limited indicator. It records legible types of economic activity: sales, corporate turnover, transactions with a market price. Everything outside the labor market—from parenting to volunteering to the informal sector—is excluded.

Rather than needs met, it counts things done. This provides us with a drastically incomplete picture of worth. Furthermore, the emphasis of GDP is on gross, the total. Negative impacts or undesirable types of growth are not deducted. When forests are cleared to make way for car parks, the construction expenditure boosts GDP.[23] We end up with counter-productive dynamics. GDP value is found in war, financial speculation, ecological destruction, climate disasters, gambling, pollution, crime, smoking, oil spills and alcoholism. According to this skewed accounting, a private doctor adds value to GDP, a state doctor draws expenses from the state, and a volunteer doctor produces zero economic value.[24]

▪ ▪ ▪ ▪ ▪

The effectiveness of material abundance and growth as correctives to social ills are increasingly seen to be questionable. Since 1980, the world economy has grown by 380 percent. Yet over those years, the absolute number of people in poverty has grown by a billion.[25]

Research into the science of happiness has also exposed a paradox: beyond a certain level of wealth, GDP growth or material wealth does not necessarily lead to greater well-being.[26] Studies across predominantly Muslim countries have found that, although the observance of Ramadan through fasting negatively affects GDP growth, individuals reported greater levels of fulfilment, happiness and life satisfaction during that time period.[27] In Costa Rica, some of the poorest populations live the longest, boosted by factors beyond income, such as the strength of their relationships.[28]

Leading economists are increasingly looking askance at the obsession with growth. The economist Raúl Prebisch, influential in shaping development debates in the 20th century, noted: "We thought that acceleration in the rate of growth would solve all problems. This was our great mistake."[29]

A new economics, attuned to the needs of the 21st century, is emerging through different traditions, from post-development to degrowth, steady-state economics to feminist economics. These strands, among other inquiries, ask us to look beyond growth to imagine the possibilities of a more prosperous and equitable economy. In their varying voices, they ask us to think critically on various questions: What do we want in our societies? What is our prosperity? How can we improve quality of life without trashing the planet? Can we exercise restraint in the economy to avoid an ecological collapse? How do we redistribute wealth and build more just economies? Rather than how much can we grow, what do we want to grow?

Addressing these questions will not be easy, for the fantasies of prosperity through more production have worked themselves into reality. Our current system has been built to make prosperity reliant on growth. Financial institutions need the growth of investments, macro-economic confidence stems from a country's prospects of growth, and social-security nets rely on the windfalls of growth.

Opening up the space for these difficult conversations begins with a wider appreciation of relations and consequences.

The true costs

Our conventional view of economics is one that rarely considers broader consequences. When we think about the economy, we usually think of products before processes, of jobs before human beings, of money before meaning.

Walk through a shopping mall or a supermarket. Every item you see will have a price tag. But that number reflects only the financial cost, a slender calculation which excludes other variables.

The price of food will not reflect the biophysical impacts of its production. The price of clothes will not include the toxicity of their dyeing. The price of a phone will not contemplate the pain of its manufacture. The price of fuel will not include the costs of fuel combustion for future generations.

When we look at the omissions of these hidden costs, we can see that markets and economic decisions are broadly oriented by one dominant type of value: financial.

Such an exclusionary vision of value yields riches in the production of costs. Businesses shift costs from their own balance sheets to society, transferring burdens—from care work to contamination—

to the least powerful groups.[30] Fortunes are made by devaluing and degrading ecosystems, territories and their inhabitants.[31]

This separation between production and its perils has crafted an economic system that is deaf to its own destruction.[32] In the bookkeeping of governments and companies, our ecosystems are footnotes. The majority of companies do not acknowledge or internalize wider costs (known as externalities) throughout their operations. One study by consultancy firm Trucost found that none of the world's top industries would be profitable if they had to fully account for their externalities.[33]

Routes ahead

The crisis of our planetary system requires significant changes in our economic metabolism, in the values that shape it and the social structures and relations that distribute its fruits.

Failing to shift course presents threats to our economic system, even on its own terms. Fishing companies face dwindling fisheries, devastated by acidifying oceans; food companies face harvests spoiled by droughts; realtors face devalued waterfront properties rendered worthless by storm surges and rising sea levels.

Just a single incident of climate violence can erode huge economic gains. In 2015, Cyclone Pam devastated Vanuatu, wiping out two-thirds of the country's economic output for that year. In 2015, tropical storm Erika caused enough damage to wipe out almost all of Dominica's annual GDP. In 2017, Hurricane Maria pummelled Dominica, causing even greater devastation.[34] In 2016, extreme weather caused property damage across India equivalent to the national health budget.[35]

Paradoxically, the policies required to shift our course also present major financial risks. The most salient of these is known as the carbon bubble. To avert catastrophic levels of climate change, most of the known reserves of fossil fuels will have to remain in the ground. But the financial value of the world's publicly traded fossil-fuel companies is largely based on the value of their proven reserves: how much they have and will extract in the future. If we want a climatically stable world, fossil-fuel companies are worth only a fraction of their public value.

A major downgrade of value in fossil fuels, and decline in consumption, would have tectonic implications across the economy, particularly for countries reliant on their income.[36] Many banks are heavily exposed to climate-vulnerable assets, as oil companies, particularly in eras of low prices, have borrowed heavily to bankroll their dividends.[37]

The magnitude of these interacting and accumulating economic risks merits intense attention.

■   ■   ■   ■   ■

Ultimately, the economy, and the ideas that shape it, is arbitrary. Poverty, abysmal inequality and ecological plunder are not inevitable fates. The laws of society are not the laws of physics, etched in stone. Although rigid and ingrained, with time and effort, they are amenable. But we have to enact bold changes—and fast.

CHAPTER 10

THE WORLD AT 1°C:
A GUIDE TO CLIMATE VIOLENCE

Humanity is something we still have to humanize.
GABRIELA MISTRAL

■ ■ ■ ■ ■ ■ ■ ■ ■ ■

When it rains everything is drenched...
but mostly the poor.
PEDRO LEMEBEL[1]

■ ■ ■ ■ ■ ■ ■ ■ ■ ■

It's not a coincidence that our land is dying
and our women are dying.
MELINA LABOUCAN-MASSIMO[2]

■ ■ ■ ■ ■ ■ ■ ■ ■ ■

The rivers have memory.
LUZ MARINA MANTILLA[3]

■ ■ ■ ■ ■ ■ ■ ■ ■ ■

And if every victim had a book,
Iraq in its entirety
would become a huge library,
impossible ever to catalogue.
MUHSIN AL-RAMLI[4]

■ ■ ■ ■ ■ ■ ■ ■ ■ ■

WHAT IS CLIMATE CHANGE doing to us? How is it rearranging our world and bearing on its life?

Before we attempt to unpack these questions, it's worth returning to the topic of humility. Climate change's petrifying power has a tendency to eclipse all else. Our terror of future possibilities and our desperation to provoke action can often lead us to see climate change in everything, ascribing a wide range of phenomena to it without clearly thinking through the links.

The common retelling of climate change contributes to this confusion. When we hear phrases such as "mega-drought," "flash flood" or "superstorm," we think of bursts of overpowering weather, engulfing helpless societies. Such depictions not only exclude the less striking forms of climate violence, but they also help to hide the multiple roots of disaster that are unrelated to the atmosphere. Ultimately, the impact of extreme weather is determined not just by its own severity, but by the conditions that weather meets on the ground.

By failing to acknowledge the messier role climate change plays, we actually underestimate its real danger, which worsens problems and renders them insoluble, rather than generating them in the first place. So as we work to understand the impacts of climate change, it is worth remembering a simple equation:

Climate violence = extreme climate conditions × social realities

Climate violence is always the result of a collision between acute weather conditions and acute social realities. Poverty, inequality, state neglect, improper planning and abandonment lay the explosives. Extreme weather lights the fuse.

Completing these variables of vulnerability can help us understand that climate change is not a discrete threat. It is the wind that blows upon all the embers that are already there, the salt that pours into our existing wounds.

The word "apocalypse" originates in the Greek *apocalypsis*: revelation.[5] Disasters reveal us. They peel away pretence, to lay bare the underlying values of society, the previous priorities of politics and the preparedness of institutions. As Didier Cherpitel, a former secretary general of the Red Cross and Red Crescent, noted: "In many cases, nature's contribution to 'natural' disasters is simply to expose the effects of deeper, structural causes."[6] Through climate violence, the memory of injustice rises to the fore.

To understand this in greater detail, let's explore the two parts of the equation.

Extreme weather and climate conditions

Monsoons, hurricanes, torrential rains, wildfires, droughts and heat waves are all part of the rhythms of our planet's meteorology. For millennia, human beings have learnt to prepare themselves for these events: the most overt expressions of nature's might. But the rise of global temperatures is increasingly affecting the expression of extreme weather.

To understand this relationship, it's more helpful to think of contribution before causation. Climate change does not create extreme weather, it aggravates it. It heightens its frequency, intensity, seasonality and reach. It boosts chances, it energizes, it strengthens, it loads the dice of probability. It increases the force of disaster. It disturbs seasons and cycles, whose balance we rely on. As atmospheric scientist Katharine Haydoe summarizes, "[c]limate change exacerbates the naturally occurring risks we already face today."[7]

Let's sketch out some of these relationships. Climate change means rising temperatures, which accelerate the process of evaporation, removing more water from soils, lakes and rivers. The atmosphere then carries higher levels of moisture: when it rains, it rains harder. When temperatures allow for snow, snowfall is intensified.[8]

Simultaneously, as the atmosphere absorbs more water from a territory, the aridity of the land increases, heightening the risk of drought and increasing the flammability of vegetation. The risk of fires igniting, spreading and destroying grows. Dried by droughts, our forests become flammable boxes of tinder.[9]

Through such processes, climate change becomes a potent disruptor of weather patterns. Rain becomes more torrential, concentrated and dispersed. Heat waves grow longer, hotter and more regular. The wind speed limits of storms are stretched. Hurricane paths are rerouted. As heat accumulates in the ocean, the temperature of seawater rises, setting the stage for stronger storms which draw on the energy of warm waters. There are also suggestions that climate change is influencing atmospheric circulation, altering jet streams. Storm systems can stall and get stuck, allowing them more time to shed precipitation in concentrated areas. These jammed or blocked storms have prolonged impacts beyond usual storms.[10]

Higher temperatures are also driving sea level rise. First, they melt ice formations across the planet, from glaciers in the Himalayas to ice shelves in the Antarctic. This melting increases the overall proportion of water held in our oceans. Second, thermal expansion, the process through which heating waters expand, generates another half of global sea level rise. When we add up these climate-driven sources, together with our increasing extraction of water from aquifers (which transfers water to the sea), we could be facing significant levels of sea level rise across this century.

Overall, the impacts we are observing, and the models they are informing, are suggesting a future of intense atmospheric aggravation.

How can we know that contemporary weather events are affected by climate change? Through a process known as event attribution, scientists calibrate how climate change may have affected the probabilities of a particular incident occurring.[11] Using computer models, researchers simulate how particular climatic events unfold. They then compare them to counterfactual simulations of that event, which explore what may have occurred if, for example, human beings had not drastically reconfigured the composition of the atmosphere. Contrasting these scenarios allows researchers to sketch out the influence. Through such analysis, human fingerprints can be found today over many contemporary episodes of extreme weather.[12]

The inequality of exposure

While the climate is being disrupted everywhere, not every eco-system is equally affected by it. After all, when we talk about a rise in temperature, we are referring to a global average. Not every place in the world will see the same temperature rise. Land temperatures are expected to warm significantly faster than the global average, as warming above oceans is tempered by the seas' absorption of heat. Certain regions, mainly those closest to the poles and tropical latitudes, are inherently more vulnerable to global warming. In the Arctic, when compared with lower latitudes, more of the energy trapped through the greenhouse effect goes into warming than evaporation, as the atmospheric layer is shallower.[13]

Natural disparities in altitude and precipitation also shape the exposure of ecosystems. Around a quarter of all the world's renewable water falls on Canada every year, while desert environments receive almost none.[14] In Nigeria, annual average rainfall in the northern Sokoto state is 600 millimeters, while along the country's southern coast, it is 3,500 millimeters.[15] Different regions, even though proximate, may have entirely different biomes.

Given such differences, geography draws the danger zones of climate violence around dry corridors, deltas, arid regions, low-lying territories and glacier-fed river basins.[16] The most severe effects will likely be felt across these areas, particularly in the tropical and sub-tropical regions of Africa, the Middle East, South Asia and Latin America.

Every region also has its own particular climatic phenomena that present risks. In Mongolia, they are *dzuds*, intense winter storms; in Morocco and Tunisia, they are *siroccos*; in the mouth of the River Plate, *sudestadas*. Then there are cyclones in the Bay of Bengal, hurricanes in the Gulf of Mexico and the *canículas* (dry spells) of Central America.

Geophysical inequalities also emerge when it comes to sea levels. Today, sea levels are rising globally at an average of 3.4 millimeters a year, a rate which has nearly tripled in the past 30 years.[17] Just as the Earth is not flat, neither is the ocean. In different areas, we see different levels of rise.[18] The West Pacific experiences rates of sea level rise three times the global average, as thermal expansion joins changes in wind and ocean currents. In Greenland, sea levels are diminishing, given their proximity to melting ice sheets.[19]

The composition of ecosystems also shapes exposure to climate violence. Built areas are usually far hotter than adjacent rural areas, as their abundant dark and impermeable materials—concrete, steel, asphalt—absorb greater amounts of heat. Cities from Santos to Skopje are already warming twice as fast as the planet, raising major fears around their possibilities to continue sweltering into the future.[20]

From a landscape's topography to the morphology of its habitats, all these factors feed into the lottery of place. But place is not destiny. It collides with social conditions.

Social conditions

Let's turn to the other element of the equation. The way an extreme weather event will unfold is determined by the economic, political and infrastructural conditions it meets.

When floods inundate communities, it is not only torrential rains that are relevant. A key part is played by all kinds of other factors: inadequate urban drainage, improper land use, absent planning, the extent of water-absorbent surfaces, the reliability of infrastructure, the enforcement of building codes, the location of toxic-waste sites, the permeability of sewage deposits, the processing of garbage and the social protection available for communities.

When landslides level communities, inadequate construction, poor road engineering, missing oversight and the clearance of erosion buffers such as forests are often to blame.

When droughts pummel communities, what is exposed is also the quality of irrigation schemes, the nature of water consumption, the strength of rural support networks and the availability of health and veterinary systems.

When lands undergo desertification, the drivers of drying are often found in cultivation practices, grazing patterns and deforestation.

When intensified forest fires raze landscapes, poor land management, inadequate responses, rural abandonment, urban migration and historic fire suppression all play their part.

When heat waves afflict populations, access to water, the intensity of labor, the strength of care networks and the availability of cooling measures all come to the fore.

In 2017, two devastating mudslides occurred in two different parts of the globe. On the evening of 1 April, torrential rains fell on the Colombian region of Putumayo, swelling the Mocoa, Mulato and Sangoyaco rivers. The expanding body of water burst through the banks, sending a sea of mud, debris and water rushing through the city of Mocoa. Seventeen neighborhoods were submerged, with that of San Miguel swept away entirely. As families slept, the torrent smashed into houses and bodies. Hundreds of people, including at least 43 children, were killed.[21]

On 13 and 14 August 2017, torrential rains lashed Freetown in Sierra Leone, collapsing an eastern hillside. Over 1,100 people were killed or rendered missing by the rushing landslide.

Although thousands of kilometers apart, the root causes of both landslides were almost identical. Inadequate urban planning, poverty, deforestation, perilous construction around water basins and failed risk management cemented the conditions. Torrential rains exposed them.

The background of the victims of both landslides also drew parallels. Many of the poorest populations of both Mocoa and Freetown were citizens displaced by civil war, forced by precarity to build their settlements on hills or riverbanks, the areas most vulnerable to debris.

What these two episodes show, along with every other incident of climate violence, is that in the throes of disaster, virtually everything is relevant. Instead of visualizing climate change as a separate threat, we need to put on the glasses of connection, and see how it interacts with the multifaceted features of our societies.

Climate violence and you

The body we are born into, the income of our family, the citizenship we are granted, the livelihood we rely on, the construction materials used in our neighborhood, the gender divisions we are socialized into, the ethnicity we are ascribed, the friends we can turn to, the size of our savings and the information we receive: all of these, and many more, are markers of fate, shaping our vulnerability to climate violence.

These distinctions—whether of gender, race, class, nationality, belonging, age, ability or caste—define our relative power within society, the resources we may have access to and the range of our available choices.

Our world is riven by deep imbalances of power, generated by historical injustices of recognition, distribution and exclusion. These imbalances attach great weight to our distinctions. We are not equal, either in our exposure to climate violence, or in our capacity to withstand it. Our experiences vary drastically.

So whenever we hear of climate change, let us think less of abstractions, and more of the intimate questions that outline the possibilities of pain.

Whose homes will resist the strength of calamities? Who knows how to swim? Who can turn on the air conditioning to allay the summer heat? Who can afford winter heating bills? Who has a car? Who can buy a bus ticket to flee an approaching hurricane? When heat waves arrive, who does intensive physical labor outdoors, and who can stay indoors? Who is housed, and who is homeless? Who has food reserves? Who lives near a hospital? Who has access to ample clean drinking water? Who is reached by warning systems? Who has access to assistance? Who is overlooked in pre-disaster planning? Who is abandoned during a flood? Who is shut out from post-disaster support? Who loses their third beach home, and who loses their family?

Who sits where in the hierarchy of care? Who is disposable?

Disasters may not discriminate but human societies do. Every storm makes landfall on a landscape riven by disparities of wealth, power and safety. And in this terrain, the law of unequal suffering holds: pain falls mostly on those most vulnerable to it. The impacts of climate change gravitate toward our fault lines. The dispossessed, the forgotten, the left behind, the frail, the malnourished and the marginalized will always pay the highest costs of disaster. Climate violence is, above all, an intensifier of injustice.

Poverties, strictures and precarities

To clarify this thought a little, try a simple thought experiment. Think of your home city, town or rural area. Now think of the most common ecological risk your neighborhood faces: is it flooding? Or landslides? Hurricanes or drought? Is it extremely hot weather or cold weather? Is it air pollution or waste? Once you've chosen a risk, try to draw a mental map of your home city: what are the areas most affected by that ecological risk?

Keep that image in the back of your mind, but this time make another map. Trace out the areas of your home area with the highest rates of poverty and deprivation. Do the two maps overlap?

Usually, the maps of climatic mortality mirror the maps of social precarity. To be poor is to be both unprotected and prone to environmental risk. Low-income neighborhoods are typically those closest to polluting facilities, dump sites and the sources of flooding.

The poorest neighborhoods are also usually those most deprived of the instruments that allow us to endure extreme weather. Risk-reducing infrastructures—from storm pipes to sanitation, from stable electricity to firm building foundations—are concentrated in the wealthiest neighborhoods. Informal settlements, on the other hand, deprived of state recognition and support, face some of the highest risks.

Wealth allows for greater choices to be made in the face of threats: you can choose where to live, what quality of water to obtain and what burdens you can avoid. In California, increasingly beset by fires, the richest members of society purchase insurance programs that provide private firefighters to protect their homes.[22] On the other hand, those with the least resources have the fewest options and footholds for recovery.

While poverty affords neglect, wealth purchases power. In the aftermath of disasters, recovery budgets, reconstruction funds and protective infrastructure are slanted to the rich. In Egyptian coastal cities, such as Dumyat and Ras El-Barr, sea defences have been set up largely to protect tourist resorts, military installations and affluent neighborhoods.[23]

■   ■   ■   ■   ■

In our world, patriarchal economic structures, power relations and entrenched expectations translate into societies with stark inequalities across genders. Women are more likely to be poor, to be denied rights, to face sexual violence, to be dispossessed of land, to be precluded from secure employment, to work longer hours, to get paid less for equal work, to be responsible for domestic food production, to be illiterate, to lack healthcare, to work closer to pollution, to be excluded from decision-making, to lack access to economic assets and to bear the burdens of unvalorized care work.

All these strictures are relevant in the context of the climate, which exposes patriarchy's fatal implications. Disasters are consistently found to be more likely to kill women than men, a disparity that is exacerbated among poorer women. In 1991, when a large-scale cyclone barrelled into Bangladesh, 90 percent of the 150,000 people killed were women. In May 2008, Cyclone Nargis made landfall on the Ayeyarwady region of Myanmar, killing 130,000 people; over three-fifths of them were female.[24] Some studies even suggest that, in disasters, the likelihood of female mortality is 14 times greater than for men.[25]

What explains these disparities? Every close detail of distinction and division gains salience.[26] During evacuation procedures, many women are homebound, tied to caring for children, elderly relatives, animals and valuables.[27] As flood waters rise, many women laden with customary heavy clothing are physically restricted from staying afloat.[28]

Broader patriarchal injustices are inflamed by disasters, which often expand the burdens of care. Called in to help in the household, young girls are more likely to drop out of schools in the wake of catastrophe.

When climate violence devastates agrarian territories, the livelihoods of women are most at risk. Women, rarely the legally recognized heads of the household, are often disqualified from the direct provision of relief, and restricted from the resources crucial for recovery: credit, insurance, employment, healthcare.[29]

In environments where women are tasked with water collection or wood gathering, climate violence can increase the physical and temporal burdens of women. Intense droughts can mean longer distances trekked to fetch water, getting up earlier to collect water or turning to unsafe water supplies.

Climate violence also augments the density of verbal, physical, sexual and emotional violence. As droughts drain water tanks, many women are no longer able to bathe in privacy.[30] In heat waves, many women risk dehydration by refraining from drinking water, in order to minimize outdoor toilet trips where they may experience harassment.[31]

In Vanuatu, reported domestic violence cases tripled after cyclones Vania and At hit the province of Tafe in 2011.[32] In Fiji, in the wake of Cyclone Winston, rates of gendered violence intensified. Shamina Ali of the Fiji Women's Crisis Center reported: "There was a lot of sexual harassment of women. There were some rapes, some reported, a lot not reported.... There were also cases of women asking for shelter and men demanding sex in return."[33]

The poverties inflicted by climate violence collide with existing precarities and norms. Rising rates of slavery, trafficking and child marriage follow extreme weather.[34] In the Sundarbans, many young women from devastated flooded communities have been forced to move into cities, often to enter the most precarious economic sectors.[35]

LGBTQI populations experience similarly oppressive strictures following disaster. Although unrelated to atmospheric conditions, the devastating December 2004 tsunami that swept across the Indian Ocean offers insightful precedent. In the wake of the devastating waves, Indian Aravanis faced systematic exclusion from relief, food provision, shelter and even death records.[36] Accounts by Aravanis who were able to gain access to emergency shelters tell of sexual and physical harassment and abuse, including "corrective rape."[37] Invisible before and after the tsunami, the majority of Aravanis were left to fend for themselves.

Gendered roles also affect men in contexts of climate violence. Debt burdens, expectations of financial responsibility and climatic shocks combine to cause disproportionate rates of male farmer suicide.[38] In the vortex of disaster, the expectation of masculine heroism also increases mortality rates. When heat waves afflict certain societies, men are more likely to be tasked with physical labor outside and left more vulnerable to extreme heat.[39]

▪   ▪   ▪   ▪   ▪

The elderly, the paralyzed, the physically challenged, those with compromised immune systems and the young also face particular vulnerabilities.[40] Impoverished health systems, the isolation of the elderly and the social exclusion of those deemed "disabled" are major risk factors for climate violence.

When Cyclone Nargis buffeted Myanmar in May 2008, many people with physical disabilities were abandoned in the rush to escape encroaching tides.[41] Nearly half of Hurricane Katrina's fatalities were people over the age of 75, with over two-thirds of the dead being over the age of 60.[42] During the European heat wave of 2003, the highest rates of mortality were seen among isolated elderly people.[43]

Evacuation procedures and early-warning systems are rarely attuned to the needs of those with physical or auditory limitations.

Power outages caused by disasters can spell death for those in hospitals, nursing homes or reliant on electric life-support systems. As roads are blocked and health centers are shut, those dependent on regular supplies of medication and those in need of operations or medical attention face particular challenges.

The type of medication we take can also expose us. Diuretics for high blood pressure can cause dehydration, whereas certain anti-psychotic drugs can contribute to preventing sweating. Studies of France's 2003 heat wave have hypothesized that many were killed by the side effects of such drugs.[44]

The bodies of the very young are also more vulnerable to the ravages of climate violence, given their vulnerability to extreme heat, early malnourishment and vector-borne illnesses.[45] Nearly 90 percent of the disease burden linked to climate change affects children younger than five.[46]

■　■　■　■　■

The imbalances of power that unevenly distribute climate risks can also be found across lines of race, clan, caste, heritage, nationality and ethnicity. In Vietnam, ethnic minorities like the Tay, Thai and Hmong peoples are disproportionately vulnerable to the ravages of disaster.[47] In Somalia, when devastating famines struck in 1992 and 2011, the majority of victims came from the minority Rahanweyn and Bantu clans.[48]

In India, Dalit and Adivasi communities have faced significant discrimination in the provision of disaster relief.[49] Rural lower-caste communities are also disproportionately vulnerable to climate-fuelled agrarian crises, having limited property access and inequitable loan arrangements. Around 70 percent of Dalits are landless farmers. The majority of farmer suicides in regions such as Tamil Nadu correspond to low-caste farmers.[50]

In Iran, minority Arab and Azeri populations face disproportionate environmental risks. The Khuzestan region, populated predominantly by Arabs, is one of the planet's most polluted areas, suffocated by chronic episodes of haze. Droughts and poorly planned hydrological projects have desiccated marshes and fields, leaving them fertile for dust storms. The region is surrounded by petrochemical factories, whose pollutants add to the noxious air.

Across Europe, Roma communities are unduly vulnerable to flooding, forced by historical oppression and exclusion into the zones of environmental risk. City councils have frequently tried to relocate Roma residents to floodplain settlements. In the Macedonian municipality of Kumanovo, Roma communities were gradually corraled into building homes on heavily polluted and flood-prone land. In January 2003, flooding swept away the homes of 406 families.[51]

In China, the *hukou* (household registration) system enshrines deep disadvantages for those registered in rural areas, who are less likely to access education, economic opportunities, subsidized housing, healthcare and welfare. These factors skew exposures and vulnerabilities to pollution and wider ecological precarity.[52]

The simple location of our birth also does much to shape our predispositions to climate change and the available support. Compare the budget of the US Federal Emergency Management Agency ($15,500 million) with that of India's equivalent authority ($100 million).[53] Or consider that the average disaster relief afforded to direct victims in the US of Hurricane Sandy ($8,608) was dozens of times the average annual wage in many other countries.[54]

Ultimately, the multiple prices of poverty explain why the poorest 50 countries in the world face just 11 percent of all environmental hazards, but over half of all associated fatalities. The richest states encounter 15 percent of hazards but suffer fewer than 2 percent of deaths.[55]

A story we can't tell

Although we may separate distinctions for the sake of clarity, in reality they are indivisible. All forms of injustice overlap and interact to generate unique experiences of injustice.[56] As Naila Kabeer writes: "While gender is never absent, it is never present in pure form. It is always interwoven with other social inequalities, such as class and race, and has to be analysed through a holistic frame-work."[57]

What climate violence leaves behind is not a set of rules, but a sea of stories.

▪ ▪ ▪ ▪ ▪

On the evening of 24 November 2012, as hundreds of workers sat behind sewing machines fulfilling orders from transnational companies, a fire broke out inside Dhaka's Tazreen garment factory. The building was a death trap: outdoor exits were missing, iron grilles barred many windows, and the adjacent road was not wide enough for firefighters to get through. Guards and managers reportedly told employees to ignore the false alarm and return to work.

The flames ripped through the building, smoke choking lungs. For workers on the upper floors, the only way out was through the windows. Guided by mobile-phone lights in the darkness, they made their way through the fumes and jumped, some to their deaths.

That night 124 people were killed. Over a hundred others sustained life-altering injuries from jumping. Many relatives were unable to identify their loved ones among the charred bodies: 53 bodies were laid to rest unclaimed.

The majority of the victims of the Tazreen crime were young women who had migrated to Dhaka, hoping to earn a living in the country's largest export industry. In research conducted after the disaster, anthropologist Mahmadul Sumon came across a startling statistic: a large number of those who died were from a small district in northern Bangladesh, notable for its water stress and depleted harvests.[58]

▪ ▪ ▪ ▪ ▪

To distill pain, we categorize. We homogenize. We count. We turn to the mathematics: digits that hold the dead, the wounded, the houses destroyed, the damages incurred.

These reductive numbers allow us to hold what we cannot: the heart-splintering volume of loss wrought by climate violence. The intimacy, the visceral imagery, the unforgettable emotional maps, of disaster.

The broken sleeps, the rotting crops, the suffocating heat, the stained liturgies, the endless scrubbing of mold. The soaked birth certificates and land titles, stained with seawater. The flooded ashes, the full morgues, the bodies unclaimed, buried without a name. The public prayers, the waterlogged pharmacies, the stuttering ventilators, the postponed operations, the endless afterlife of disaster.

The saline lands, the unrecognizable territory, the scarred earth, the brackish water, the untenable house. The resignation of uncertainty, the sadness, felt in the body, in the breath, in the words. The debt payments, the squandered savings. The slow violence of malnutrition, the shattering of prospects, the confusion of loss, the traumas we can't conceive.[59]

■   ■   ■   ■   ■

Privilege represents our shelter from the effects of a particular injustice, our distance from a deprivation.[60] Unless we are attentive to our distance, we can grossly misread the urgency of an issue, confusing our own perception of risk with that borne by another.

As we work to understand the future of climate change and what kind of response we should configure, let us root our urgency as best as we can in the realities of those most vulnerable to the rigors of atmospheric pain.

Let us also expand the scope of our focus beyond what we consider to be "environmental." As we have seen, everything about a society is relevant in the face of climate violence: its readiness, its wealth, its cohesion, its levels of equality, its political responsiveness and its treatment of its most vulnerable groups.

If we acknowledge these variables, the upshot is that when we talk about tackling climate change, we mean two fundamental things: tackling the root causes and tackling the deficiencies that makes us vulnerable to its impacts. Just as we must combat the separation of humanity from nature, we must combat the separation of climate change from the deprivations it deepens.

Climate change is a problem of colossal size, because it is a problem that multiplies. We need to retrace our maps along the lines of the new magnitude. The solution is not just a reduction of emissions—it is rather a reduction of injustices. Every gradient of warming we fail to slow needs to be compensated by an equal reduction in deprivation.

FUTURE

We are at the crossroads now.
We either say: this thing is too big for us, this task cannot be done.
[Then] we will be transformed by nature, because we will end up
with a planet warming by 4, 5, 6 or even 12 degrees.
It would be the end of the world as we know it....
Or we say: We're doing the transformation ourselves.

HANS JOACHIM SCHELLNHUBER,
climate scientist, Potsdam Institute[1]

■ ■ ■ ■ ■ ■ ■ ■ ■ ■ ■

There are no realistic solutions to the current planetary crisis. None.
A peaceful, just-in-time transition toward low-carbon, rationally
regulated state capitalism is about as likely as a spontaneous
connecting-the-dots of neighborhood anarchism across the world.
Simply extrapolating from the present balance of forces,
one most likely arrives at an equilibrium of triaged barbarism,
founded on the extinction of the poorest part of humanity.

MIKE DAVIS[2]

■ ■ ■ ■ ■ ■ ■ ■ ■ ■ ■

CHAPTER 11

A PLAUSIBLE FUTURE:
APPROACHING APOCALYPSE

WHAT DOES THE WORLD of tomorrow look like? Much of our environmental imagination is unhelpfully borrowed from Hollywood, where extravagant events lead to inconceivable misery. In *Waterworld*, climate change has melted the polar ice caps, submerging the world in an ocean of despair. In *The Day After Tomorrow*, the city of New York is consumed by an ice storm.

Disaster books and films may be captivating, but their portentous imagery misleads us. We assume all climate change will be spectacular, and since the spectacular is unlikely, the impacts of climate change must be unlikely. We envision a simultaneous Armageddon sweeping across the planet. But climate violence is not an apocalypse for all. It is one of reinforced injustices and deeply unequal fates.[3]

Some suggest that, at some point, the signals will be so clear, the alarm bells so shrill, that humanity will have no choice but to act. As conditions worsen, our resolve will strengthen. But it is unlikely that we will ever get there.

The Apocalypse will never feel like one. Instead, the world will end as it does for many today: silently, callously, unheard of.

There will be little public tragedy. Politicians will not arrive at the bedside of victims. Days of mourning will not be declared. Newspaper columns will not stretch to cover the fallen. What won't bleed won't lead.

Trendlines

What plausible shape will our world take? Future environments will be defined by intense instability. All cities and rural areas will face rising climate risks, including droughts, heat waves, forest fires and floods. Parts of the Middle East and South Asia are likely to be uninhabitable within the next decades, as precipitation falls and temperatures spiral. The borders of the world will likely be redrawn into spaces of sanctuary and flight.

As the pace of sea level rise accelerates, the world will be afflicted by what is known in Kiribati as *baki-aba*, land hunger. Human life will be confined to smaller stretches, as inland territories desiccate and coastlines are shaved by encroaching waters. Aquifers and croplands will be increasingly corroded and contaminated by saltwater. Overreaching tides will lap into roads and homes.

The temperatures we have locked into warming already have set in motion the loss of major ice sheets. Some estimates have warned that failures to divert our course could lead to average sea level rises of 30 centimeters by 2050, and over two meters by 2100.[4] If emissions continue at current rates, Antarctica alone has the potential to add a meter of sea level rise by 2100.[5]

In this new era, what is low will go. Low-lying states such as Vietnam, the Netherlands and Bangladesh. Densely populated littoral areas such as New South Wales. Cities such as Cairo, Kolkata, Tianjin, Osaka, Bangkok, Rio de Janeiro and Kinshasa.

Riverine deltas by the Nile, Mekong, Moulouya, Mackenzie, Cauvery, Paraná and Yellow rivers, highly fertile and vital for food security, will face submersion. River basins around the Paraguay, Amazon and Limpopo rivers will experience severe water stress. Many regions reliant on highly vulnerable aquifers will become unable to support agriculture or stable human life.[6]

Entire island archipelagos are likely to be swallowed by the waves, including Kerkennah (Tunisia), Guna Yala (Panama), the Solomon Islands, Fiji, Cape Verde, the Maldives, the Maluku Islands (Indonesia), the Torres Strait Islands, the Seychelles, Micronesia, Kiribati, the Bahamas, the Moonsund Archipelago (Estonia), the

Lakshadweep chain and Severnaya Zemlya (Russia). Their uprooted peoples will be destined to be diasporas. Coastlines will be washed away. Some flags will never fly again on the land they represent. Maps will be annually outdated, as history textbooks are rewritten.

The survival of entire communities, cities, valleys, countries and regions hinges on the rapid reduction of emissions. Unless such cuts are made, crisis may become chronic. Recovery will be impossible, if not a way of life, for communities caught in the never-ending labor of salvage. Inequalities will grow intractable, as calamity after calamity depletes resources and energies. Emergency will be rendered permanent—a constant reality of droughts, floods, plagues, epidemics and heat waves. Many cities will have to assume a cycle of recurrent evacuation, cleaning, reparation and repopulation.

For ecosystems, the gaps between droughts, furious storms and bleaching events will continue to shrink. The intervals required for repair and regeneration will shorten into absence, as fire seasons stretch into permanency. Lands pummelled by extreme weather will lack the time to heal.

The pace of relentless destabilization will exceed the pace of adaptation. The consequences will be devastating for those directly reliant on the sustainability of lands; after major losses from drought, it can take pastoralists decades to rebuild their herds.[7]

▪   ▪   ▪   ▪   ▪

While models may project the future and sketch out its possibilities, any exactitude is impossible. We are heading into the great unknown: oceanic and atmospheric conditions unprecedented in the last 20 million years.[8]

Climate change represents the extinction of predictability, a demolition of certainties. By the end of the 21st century, the environmental conditions on up to a third of the Earth's surface will be simply unprecedented for presently living organisms.[9]

Our societies have evolved on the basis of observation and adjustment. Cultural and economic practices have been tuned to the climate of our histories. Yet, subject to the confusion of ecological precarity, the compasses of the past lose their power of guidance. Traditional rain calendars are rendered inaccurate. In villages of the Indian state of Odisha, people remember six seasons: *grishma* (summer), *barsa* (monsoon), *sarata* (autumn), *hemanta* (pre-winter), *shishira* (winter) and *basanta* (spring). Today they have blended into two: the rains and the summer.[10]

Climate violence involves the theft of the future, the slow dissolution of people's ability to shape their tomorrows. Previously assumed stabilities are rendered conditional. This spells anguish for communities reliant on the consistencies of their environment. For most smallholder farmers, the margins of survival are minimal. Unreliable rains mean unreliable incomes, unreliable currents mean unreliable fish catches. In the Siberian Arctic, reindeer herders are dependent on stable ice floes, and on summer rains to moisturize the tundra and allow for sleighs to be drawn.[11]

Today's events are merely the early stages of climatic breakdown. There will be no "new normal." We will constantly be testing and pushing back the contours of possibility. The unprecedented will become expected. Events that formerly occurred once in a thousand years will become recurring. New meteorological categories will emerge to capture runaway realities. Aberrations will impose themselves as permanent features. Ranges will grow unthinkable, as averages are pulled upward. The parameters of necessary action will change: from 1.5°C, to 2°C, to 2.5°C, to even trying to stay below 3°C. The rules of the game will keep having to be rewritten.[12]

In August 2017, Houston was lashed by Hurricane Harvey. Some of its neighborhoods received over 50 centimeters of rain in a matter of hours. The National Weather Service pronounced: "All impacts are unknown and beyond anything experienced."[13]

Slowly, it is the past that is becoming abnormal. In December 2017, the database of the US scientific agency responsible for charting atmospheric conditions (NOAA) automatically flagged and removed unprecedented temperature rises in the Alaskan town of Utqiag.vik, considering them as impossible outliers.[14]

Biomes will shift, as boreal forests turn into steppes, and tropical forests dry into savannahs. The Arctic will be a "New Arctic," shedding its icy state and frozen past. Some areas of the planet will face year-round fire regimes.[15]

Ecosystems will be gradually eviscerated, with coral reefs perhaps the first to disappear. The subnivium, the seasonal refuge of life that forms beneath the snow, faces major threats.[16] Sea-feeding rivers will struggle to reach the coastline. Water bodies, such as Lake Chad, Lake Urmia and the Caspian Sea, will continue to dry and diminish.

Health

Public-health researchers have also described climate change as the prime threat to global health this century, given its potential to taint freshwater bodies, extend periods of hunger and expand the scope of disease infection.[17] Many vectors for disease thrive in warmer climates, and climate change has allowed for the spawn of climate-sensitive diseases such as dengue, malaria, chikungunya and Japanese encephalitis.

Various diseases long thought to have been eradicated are locked in carcasses frozen beneath permafrost. With climate change melting the frozen ground, fears abound that thousands of pathogens could be released. In 2016, anthrax broke out in the Siberian Yamal Peninsula, when a heat wave thawed the permafrost, allowing for spores to surface from old frozen carcasses.[18]

In mountainous regions, melting glaciers are releasing old toxic pollutants, such as DDT and polychlorinated biphenyls, that were banned decades ago. The gradually disintegrating permafrost also contains colossal deposits of mercury, which risk entering environments and cascading across ecosystems.[19]

The risks of climate change extend to mental health, as extreme weather can involve extreme shocks to our being. To thrive, human beings require a degree of stability and certainty; empirical studies show the tight links between climate safety and happiness.[20] But as stabilities are shaken, climate violence is already exacting a profound psychic toll the world over, leaving swelling rates of loss, anguish and suicide in its wake.

As psychiatrist Lise Van Susteren explains: "When the place you call home is burned down, blown away, dried up or flooded; when you lose your possessions or your pets, your livelihood, your community; and see injuries, illness and death, the mix of fear, anger, sorrow and trauma can easily send a person to the breaking point."[21]

Iowan farmer Matthew Russell, whose family has tended to its land for five generations, recounts: "Psychologically...there's a lot of anxiety that I don't remember having 10 years ago...there's this tremendous anxiety around the weather because windows of time for quality crop growth are very narrow."[22]

Climate violence does not cause mental illness; rather, it exacerbates existing conditions, and adds to the devastating cocktails of factors that drive so many to misery. In India's rural communities, climate shocks afflict farmers that are already facing mounting debts, the rising costs of agricultural inputs, water shortages and volatile prices.[23] Since 1995, over 300,000 Indian farmers have taken their own lives. In 2015, a farmer committed suicide every hour.[24] In Maharashtra, pummelled by drought, 2,414 farmer suicides were reported in the first 10 months of 2015.[25] Behind every farmer accounted for in statistics are invisible stricken families, grief-consumed colleagues and traumatized children.[26]

In the Indian states of Tamil Nadu and Karnataka, hundreds of farmers have died of heart attacks associated with environmental stress. Cardiologist R Bharathiselvan reflected:

> I have never experienced anything like the mortality rates that have happened...It's the stress of not being able to harvest a crop. It's the feeling so many farmers have that there is no other path for them to follow. They suffer between denial and dependence and see there is no escape route.[27]

For communities intimately connected with their land base, climate violence directly assaults the territory with which they are intertwined. The Inuit are among many Arctic communities experiencing *uggianaqtuq*, the strange behavior of the environment. Shrinking ice, shortening winters and rapidly changing weathers have been linked to widespread anxiety, confusion and mental strain. As one community member noted to the public-health researcher Ashlee Cunsolo: "Inuit are people of the sea ice. If there is no more sea ice, how can we be people of the sea ice?"[28]

New horizons of heat

Climate change will also increase the likelihood and severity of heat waves, the deadliest form of disaster in many countries. Nearly half of the world will soon face regular deadly heat waves.[29]

Nine out of history's ten most fatal heat waves have occurred in the past two decades, killing over 128,000 people.[30] In the last decade alone, the loss of human life from heat waves has spiralled by 2,300 percent.[31]

In 2015, thousands of people were killed across India and Pakistan, as heat waves scorched communities. The following year, some Pakistani cities dug mass graves pre-emptively to accommodate the expected victims of summer heat.

New temperatures will test the limits of human endurance. Scientists have researched what is known as the "wet-bulb threshold," a dangerous combination of humidity and heat. This "survivability threshold" is breached when temperatures surpassing 35°C combine with humidity levels above 90 percent.[32] Beyond this level, the body cannot cool itself through sweating. When exposed to severe heat, the body forces blood toward the skin, straining the heart. Two conditions emerge: ischemia and damaging heat cytoxicity. Under this kind of heat stress, cells deteriorate and organs are damaged.

Unsurvivable climatic conditions will gain prevalence. Across the Arabian Gulf, we can expect heat waves that create conditions beyond the boundaries of human survival.[33] Under current trends the Hajj, the sacred pilgrimage of Islam, could become physically impossible to undertake in summer seasons.[34]

Recent heat waves, even in some of the world's richest states, illustrate the stark stakes. A two-week heat wave killed over 35,000 people in Europe in 2003, as forest fires flared across several countries and rivers dropped to rock-bottom levels. The heat wave was designated a "once-in-500-year event"; it recurred three years later.[35]

In the summer of 2010, a potent heat wave swept across Russia; Moscow's extreme heat record was equalled or exceeded five times in a two-week period.[36] Around 56,000 people were killed by the heat and the resulting air pollution.[37] Over a third of Russia's wheat crops were lost, contributing to rising grain prices and unrest across the globe.

As geographer Camilo Mora describes, "for heat waves, our options are now between bad or terrible."[38] Even in the best-case scenarios, almost half of humanity will face deadly heat by the end of the century.

Temperatures in the Cauvery basin, India's rice bowl, are forecast to rise by almost 4°C by 2080, a level which will likely translate into a 40 percent collapse in grain yields.[39] As C. Subramaniam, a farmers' leader in the village of Vettaikaraniruppu, reflects: "We used to easily be able to harvest three crops a year. Now we can barely harvest a single crop. Forget water for farming. We don't even have water for drinking." As extreme heat strains food production, the pain is compounded by the loss of land; over the past four decades, the Cauvery delta has shrunk by a fifth.[40]

As temperatures rise, seas absorb coastlines, farmlands are depleted, populations grow, and resources gain scarcity, contests over land are likely to inflame. Many island nations have turned to reclaiming land from the sea, importing huge amounts of sand, often obtained through the destruction of river ecosystems or other islands.[41] Countries are increasingly competing over access to resources. Resource-grabbing, a phenomenon where communities are dispossessed of their access to land, marine and natural resources by commercial interests, is on the rise. Many investment banks, sovereign wealth funds and state authorities are placing long leases on farmland. Countries such as South Korea, Saudi Arabia and Qatar have secured large tracts of land in Africa.

Water will also find itself at the center of struggles, as climate change saps at reserves already under threat by human consumption. Over recent decades, we have poisoned, depleted, dammed and drained our water bodies at intensive rates. In the last 50 years, freshwater extraction has tripled and the extent of irrigated land has doubled.[42] Half of the world's largest aquifers are being depleted, with a third in a condition of serious distress. A fifth of rivers no longer reach oceans. Many states are committing "hydrological suicide," significantly overexploiting their own water sources. These trends risk augmenting existing precarities, with around 60 percent of the world population living in areas afflicted by water stress and shortage.

The aquifers that nourish the North China Plain, home to 200 million people and most of China's wheat production, may dry within decades. Many Himalayan glaciers, the water towers of Asia, are retreating at a pace of 20 meters a year, and face complete evaporation within decades. Some of the world's most important rivers—the Mekong, the Yangtze, the Salween, the Brahmaputra, the Ganges, the Indus and the Yellow River—are fed by these Himalayan glaciers and glacial lakes.[43]

The Fertile Crescent, the birthplace of human agriculture, may lose all its fertility by the end of our century.[44] Without a severe reversal of emission trends, large areas of the mid-latitudes (southern Africa, South Asia, Western Asia, Central America, the Sahel, the Maghreb) will see routine drought and agrarian devastation.[45] IPCC reports have outlined that humanity may encounter "a breakdown of food systems linked to warming, drought, flooding, and precipitation variability and extremes."[46]

As some areas dry, others will bear the weight of storms. Water, in its excesses and scarcities, will be definitive, with 9 in 10 of the most extreme weather events related to water.[47] There are three different types of flooding: coastal, fluvial and pluvial flooding. Climate change, coupled with increased groundwater usage and damming, risks augmenting all three.

Adaptation and loss

Major climatic impacts are inevitable, yet we are grossly unprepared. Early-warning systems, vital mechanisms to protect populations, are missing in over 100 countries.[48] The Red Cross has urged greater investment in disaster risk reduction before it is "too late for too many."[49]

Yet our capacity to adjust is not infinite. As the IPCC's tactful language outlines, we have "limited potential for adaptation."[50] When those limits kick in, we face the violence of loss. The disappearance of coastal cultures, of island life, of riverine folklore. The erosion of homes, customs, languages, stories and ways of being. The end of communities as they know themselves, as with the Uru-Murato people, forced to abandon their home around Bolivia's disappeared Lake Poopó, and the Biloxi-Chitimacha-Choctaw people, who have lost 98 percent of their land in Louisiana, US, to erosion and rising sea levels since 1955.[51]

Every day, our expanding memorial of climate violence adds the names of gradually abandoned places: Shishmaref, Taro, Kivalina, the Carteret Islands, Kepidau en Pehleng, Nahlapenlohd, Parali, Mundrothuruthu, Tegua. Every day, homes, cemeteries and archaeological sites are washed into the ocean.

As much as we may want to deny it, or coat it in euphemisms such as "strategic withdrawal," the inevitable future involves a huge retreat from coastal, riverine and arid regions.[52] Displacement will arise as territories turn into deserts, family homes corrode into the sea, and crops wilt in salinized soil. Over 20 million people are already displaced each year by disasters, at a rate of one every second.[53] Two million people have already been relocated and resettled in population transfer programmes.[54]

Dozens of communities, from the Fijian village of Vunidogoloa, to the Senegalese city of Saint-Louis, to the Cuban fishing community of Palmarito, are having to pack up and move.[55] The island nation of Kiribati has bought 2,500 hectares of land in Fiji, a purchase intended to provide a home for future refugees. In Panama, the Guna inhabitants of the Gardi Sugdub island have acquired various mainland hectares to establish a new village community.[56]

Some will pay their way to safety. But those who cannot afford to go will likely be trapped by their own poverty. It costs money to migrate. One study in Malawi found that "climate change is likely to increase barriers to migration rather than increasing migration flows."[57] Furthermore, many people are understandably reluctant to abandon their homes, their friends and the burial grounds of their ancestors.[58]

Yet neither does relocation guarantee safety. Dubuc, a settlement expanded to house Dominican citizens displaced by Tropical Storm Erika, faced the devastation of Hurricane Maria just two years later.[59] Many of the areas already serving as sanctuaries for displaced peoples are facing increasing climatic strains. Jordan, the state with the second-highest number of refugees per capita, will potentially see droughts double by 2100 and average temperatures rise by 4.5°C.[60]

In this future of climate crisis, what will remain of the right to flee? Borders may be thinning for pollution, but for people, they will almost certainly thicken. Rather than regulating the drivers of displacement, authorities are likely to regulate movement even more severely. Those embarking on perilous sea and land journeys today meet the brutalities of borders and bureaucracy, a world of fortified walls. When the Berlin Wall was demolished in 1989, there were just 15 border walls across the world. Today there are 70.[61]

In the last decades, tens of thousands of human beings seeking better lives—the equivalent of several sunken *Titanic* cruiseliners—have died trying to reach the shores of safety. The Mediterranean, Arafura and Caribbean seas have turned into cemeteries of water.

And in those states where the displaced arrive, as author Robtel Neajai Pailey notes, "undocumented migrants" are becoming "the

scapegoats of politicians as a cover-up for their failures in respond-
ing to the needs of the documented."[62]

As climatic violence continues to displace communities, how
will politicians react? How will climate change affect the ways in
which politicians mobilize our basic desire for safety? What human
furies might the atmosphere unleash?

Blame and opportunism

Across human history, the force of climatic disaster has always
raised the question of blame. Amid the wreckage of storms and
droughts, societies have pointed their fingers at the rage of the
gods, at the natural cycles of life or at their own sins.

While these arguments are less relevant today, as deprivations
compound, climate violence will become more relevant in politics,
feeding into political dispute and confrontation.

The uncertainty of the climate allows for flexible blame. Climate change presents the perfect cloak, a way for obscuring any
political or social reasons for a population's vulnerability. Political
elites can use environmental shocks to divert attention away from
state neglect, or to conceal their own agendas behind environmental imperatives.

The potency of climate violence hands politicians a powerful
tool to launch policy in the name of ecology or safety. Already land
grabs have been justified in the language of sustainability, conservation and food security. In the Maldives, resettlement projects
that sought to consolidate a population stretched over 200 islands
on just a dozen were long proposed by the state for economic and
political reasons. Yet today, these projects, previously considered
deeply unpalatable by citizens, have acquired momentum after
being framed as necessary environmental measures.[63]

In China, most of the populations resettled in the Ningxia region through ecological transfer programmes are Hui Muslim.
These targeted resettlement policies are rumoured to be linked to
the control of ethnic-minority populations, with environmentalism
as cover.[64]

In the Colombian province of La Guajira, thousands of children have died in recent years from thirst and malnutrition. State organs and politicians have been quick to invoke "climate change" as a cause behind these deaths. While changing rain patterns were certainly at play, the central culprit was the state. Years of water-intensive coal mining, corruption, racism and institutionalized neglect assembled the conditions for such human rights violations to occur.[65]

Political forces could also take advantage of environmental crises to arouse prejudices, misroute grievances, enrich elites or enshrine draconian policies.[66]

Moments of ecological precarity are ultimately moments for power. In 1930, Rafael Trujillo became President of the Dominican Republic, following a coup. Two weeks after his inauguration, the San Zénon hurricane lashed the country. Capital city Santo Domingo was strewn in debris, with 90 percent of its buildings destroyed; 4,000 people were killed. The disaster proved to be a perfect opportunity for Trujillo to solidify his status through the myth-making of reconstruction. A raft of emergency laws was passed to hand the leader dictatorial powers. Santo Domingo became Ciudad Trujillo.[67]

Authoritarian and prejudicial powers have similarly drawn strength from scarcity. Nazi officials deployed arguments of scarcity while seizing the fertile lands of Ukraine, eliminating Jews and securing European resources for the German people.[68] In 1900, a hurricane struck the Texan town of Galveston. Major water surges inundated the town, killing 8,000 people. Dozens of black people, accused of looting, were executed in its wake.[69]

The destructive fusion of ecological dread, imperialism and scapegoating could see new incarnation, as states deploy climatic arguments to secure standards of living and codify scapegoats. We could see defence recast as ecology, exclusion as justice. The demagogues of a warming world will ride on crisis, convincing others that our societies are polluted by resource-draining immigrants, or deploying narratives that warn of overpopulation.[70]

The language of ecology is easily twisted into the language of hatred. Far-right movements frequently invoke environmental metaphors to demonize and dehumanize others. Those striving for better lives become an "avalanche," a "flood," a "swarm" of migrants. Movements of human beings become waves, tides and deluges. For centuries, elites have used the discourse of "cleanliness" to cast the less powerful as culprits of contamination.[71] Jews, Roma, Chinese immigrants and other groups have all been portrayed as invasive species, vermin, viruses and germs. Aspirations to restore the "natural order" have been used to strip others of rights and to reinforce unjust fictions of gender and race.

Without care, environmental concerns can be warped and forged into justifications for xenophobia. Anti-immigration groups in various countries have supported environmental causes, tying natural conservation to population control, secured by restricting the entry of foreigners.[72] Visions of organic and local food have been used by extreme nationalist groups to advocate visions of purity around the land and its races.[73]

Ecological conflict

Today, land and the waters to nourish it are at the root of many conflicts around the world. One common fear around climate change is that its disruption of weather patterns and resource availabilities could fan the flames of violence, triggering paroxysms of brutality.

Various studies have suggested links between climate change and rises in domestic violence during droughts, urban violence during heat waves, and civil wars.[74] Environmental pressures have also been identified as potential contributors to brutal episodes of atrocity, from the Rwandan genocide to the war in Darfur. Many contemporary armed groups have been able to find recruits in agrarian communities ravaged by crop failure. From Iraq to Niger, in drought-stricken areas, militants have arisen as benefactors and employers, handing food to stricken farmers and offering jobs to unemployed youth.[75]

This research suggests causal links between climate change, aridity, crop failure, livelihood degradation, pastoral mobility and the physiological effects of extreme weather.[76] Violence is particularly likely in the absence of communal institutions that can manage conflicts.[77] In 2016, after droughts and subsequent crop failures, thousands of peasant farmers in the Philippines blockaded roads, demanding food relief from the government. After a few days, the state responded: its police and armed forces shooting at the blockading crowd. Ten people were killed and 87 went missing.[78]

Violence arises not necessarily in those areas most beset by scarcity, but in areas with greater relative amounts of resources.[79] Near Lake Turkana, the world's largest desert lake, Turkana and Dassanach communities have found themselves embroiled in confrontations following chronic drought. With temperatures swelling, and development projects tapping dwindling water reserves, thirst-stricken livestock were being lost at alarming rates. Traditionally pastoralist Turkana families were pushed towards fishing livelihoods; now they were competing over access to land and water with the Dassanach. In 2011, 65 people were massacred in the Turkana village of Todoyang; 100 Dassanach people were killed soon afterwards.[80]

But we must be careful not to reduce the convolutions of violence to the simple mechanical outcomes of an overheated atmosphere. In certain cases, the deprivations wrought by climate change can reduce conflict, as people band together to manage scarce resources. Studies of drylands in Kenya's Marsabit District found that violence actually diminished during drought.[81]

Subsuming conflicts under the linear trajectory of climate change can do much to bleach the ample roots of violence. The violence of Syria has been particularly prone to crude ecological explanations. Former US vice-president Al Gore observed that the "underlying story of what caused the gates of hell to open in Syria" was a "climate-related drought."[82] According to this narrative, between 2006 and 2010, droughts across Syria devastated agrarian communities, driving millions into extreme poverty and displacing hundreds of thousands into urban centers.

This simple story has two major flaws. First, the strength of Syria's drought did not stem from climate change alone. The reckless expansion of industrial herding, overgrazing, the intensified cultivation of water-intensive crops, the proliferation of water drilling, the corrupt use of irrigation funds and the shredding of rural support networks were processes that over decades laid the foundations of vulnerability to drought.[83] As Francesca de Châtel, a researcher of Middle Eastern hydrology, explains, "[o]verstating the importance [of climate change in the Syrian context]," is an "unhelpful distraction that diverts attention away from the core problem: the long-term mismanagement of natural resources."[84]

Second, ascribing Syria's conflict to climate change neglects the prime cause of the uprising: the violent, repressive and kleptocratic legacies of the Assad dynasty. In September 2016, dozens of Syria's leading intellectuals and regime opponents signed an open letter condemning US and Russian policies in Syria. In it, they also criticized conventional interpretations of the conflict:

> Many, especially in the West, prefer to hide behind fatalistic theories steeped in religion or culture—when they do not attribute events to climate change. This explains why a bad situation has become much worse, but it also absolves the powerful elite, including Bashar al-Assad and his gang, of their political responsibilities.[85]

To navigate the world of tomorrow, we need to begin to talk more responsibly about climate change.

A world beyond 4°C

The best-case scenarios for climate change ensure extensive suffering. The worst-case scenarios guarantee outcomes beyond the scope of our imagination. If we fail to act swiftly and seriously, we simply lack the ability to depict the depth of possible change. All we have is the lifeless language of projection.

If the feedbacks of the carbon cycle are strong, 4°C of warming could be reached within 40 years.[86] According to the World Bank, a 4°C world would be "incompatible with organized global community." In the diplomatic language of the IPCC, 4°C means "severe, widespread and irreversible impacts" for billions.[87] Climatologist Rachel Warren notes that "in such a 4°C world...the ecosystem services upon which human livelihoods depend would not be preserved."[88]

Military planners also outline their view of tomorrow: "outright chaos."[89] In 2003, a Pentagon-commissioned report outlined that the "United States and Australia are likely to build defensive fortresses around their countries because they have the resources and reserves to achieve self-sufficiency.... Borders will be strengthened around the country to hold back unwanted starving immigrants from the Caribbean islands (an especially severe problem), Mexico and South America."[90]

Four years later, another report from a Pentagon-linked research group analysed Africa, Asia and the Middle East:

> Many governments in the region are on edge in terms of their ability to provide basic needs: food, water, shelter and stability. Projected climate change will exacerbate the problems in these regions and add to the problems of effective governance...Economic and environmental conditions in these already fragile areas will further erode as food production declines, diseases increase, clean water becomes increasingly scarce, and populations migrate in search of resources.[91]

These projections may seem far-fetched, but many governments, militaries and corporations are planning for four degrees of warming. "Planning" may suggest that four degrees will be somehow manageable, when it will be anything but. The last times our Earth encountered comparable greenhouse-gas concentrations and emerging climatic circumstances, the poles were ice-free and sea levels were dozens of meters higher.[92]

Reactions and responses

All we can be certain of is that tomorrow will be unrecognizable and unpredictable. But what is likely? The fractures of our contemporary world answer that question for us.

If we were heading for catastrophic warming as a united global society, guided by strong convictions of solidarity and compassion, that would be one thing. But we are sailing towards it as a divided world—each for themselves, every state for its own. Under such conditions, our world afflicted by climate violence will resemble an autarky of ambivalence. The poor will be left to doom, enclosed into reservations of merciless misery. Walls will be fortified to keep out the undesirable, as those elites hemmed into oases of safety use violence to ensure their own protection. Climate change will be a death warrant for those unable to afford distance from its impacts.

And although we may focus on decades, perhaps even on centuries, it is likely that the changes unleashed by our warming of the planet will transcend millennia. Recent research suggests that our accelerating trends of sea level rise may be unstoppable for thousands of years, elevating ocean levels by as much as 50 meters.[93]

So the question becomes not, how can we save the world, but rather, how do we save each other?

How do we assemble the difficult utopia of climate safety? What will be the ethical identity of tomorrow? Who will have the right to survive? How do we grant people humanity in our coming age? How do we tame and replace our destructive methods of development? How do we build an economy that abandons its obsession with lifeless indicators, and values the productivity of dignity and justice? What boundaries does nature place on our own?

CHAPTER 12

A POSSIBLE FUTURE:
THE WORLD WE CAN WIN

Ar scáth a chéile a mhaireann na daoine.
[It is in the shelter of each other that the people live.]
Irish saying

■ ■ ■ ■ ■ ■ ■ ■ ■ ■

Because many of our dreams were reduced to what exists,
and what exists many times is a nightmare, to be utopian is the most
consistent way of being a realist at the beginning of the 21st century.
BOAVENTURA DE SOUSA SANTOS[1]

■ ■ ■ ■ ■ ■ ■ ■ ■ ■

Ecology is not "love of nature," but the need for self-limitation
(which is true freedom) of human beings...
CORNELIUS CASTORIADIS[2]

■ ■ ■ ■ ■ ■ ■ ■ ■

We live in capitalism. Its power seems inescapable.
So did the divine right of kings.
URSULA LE GUIN

■ ■ ■ ■ ■ ■ ■ ■ ■

The right to be happy is very subversive.
That's why we should aspire to be happy.
BERTA CÁCERES[3]

■ ■ ■ ■ ■ ■ ■ ■ ■

POLITICS IS ABOUT the transformation of reality. Human beings have always fought to mould the present, to assuage the difficulties of life, to break open freedoms and to make other futures possible. Social movements have overthrown empires, toppled dictatorships and shaken the foundations of hierarchies. Public healthcare, the weekend, the eight-hour day, paid holidays, universal suffrage, labelling on food, the end of child labor, the formal eradication of slavery, autonomous indigenous territories, the delegitimization of homophobia: all these are rights that have been fought for, not gifted.

This inheritance of memory demonstrates that despair is always defiable. The first step toward confronting our ecological crisis is to recall our potential and to explode the myth of impotence. For too long, our politics has been evacuated of utopias, our dreams carved into doubts, our possibilities circumscribed to probabilities.

Dominant ideology hands us a paradox: it promises us infinite choice yet declares there is no real alternative to our current system. We can dream of a little more individual wealth, a little more social good, a little more distribution, but not more. Our menu of alternatives involves a small selection of policy tweaks and market-based solutions. Envisioning anything broader feels futile or impossible.[4]

Immersed in its uniformity, we are unable to look at the world askance, and consider how things could be if they were organized differently. As philosopher Roberto Mangabeira Unger asserts, "[a]t every level the greatest obstacle to transforming the world is that we lack the clarity and imagination to conceive that it could be different."[5]

What surrounds us is ultimately a story—a very powerful story, but ultimately one that is replaceable. Everything can be changed. Few things are inevitable. The present that surrounds us is the outcome of contingent situations. Our political system, the economy, the dominant culture are in fact lived fictions, momentary products of the past. The world is made of clay, disguised as granite. We need to break through the confines of our imagination.

What if the energies devoted to needless marketing were devoted to education? What if the fortunes expended on perpetuating violence were redirected to public health? What if we had an economy based on the productivity of justice, the creativity of culture and the dignity of life? What if the balance of the natural systems that support life were as important as the balance of financial books?

The most fundamental change we need to make is a radical shift in paradigms from the logics of simplicity, control, uniformity and separation to those of diversity, complexity, care and connection. We must shed the inheritances of our dominant vision of prosperity, rooted in human superiority, masculinist values, state power, individualism and consumerism. The strength of our societies needs to be dissociated from our power over nature.

From a climatic perspective, our transformation requires two axes: a massive reduction of emissions and a massive reduction of injustice. We must diminish vulnerabilities, reduce the drivers of disaster and open room for well-being, restoring ecosystems and human relationships.

The singularities of modernity should give way to pluralities. Today, the monocultures in our fields mirror the monocultures in our societies, visions and solutions. These have wrought a mono-world, at odds with our intricate and pluralistic universe.[6]

As the Zapatistas and others in their wake have demanded, we need a world where there is room for many worlds. Through such diversity, and a "dialogue of knowledges" across societies, we can learn from each other and craft collective visions that bring together the best ideas and practices of the world's peoples.

Anthropologist Arturo Escobar called this "the pluriverse": a polyglot kaleidoscope of ideas. All the various cultures of our world are inimitable refractions of the human spirit, a vast repertoire of values that can assist us in creating the necessary societies of an ecological future.[7]

These visions and ideals include: *buen vivir* (Spanish, good life); *utz' k'aslemal* (Maya, harmony with nature); *swadeshi* (Hindi/Sanskrit, local production); *ubuntu* (Swahili, human kindness); *ukama* (Shona, a state of affinity with others and the cosmos); *kapwa* (Tagalog, shared being); *shi-shi wu-ai* (Chinese, a harmonious relation of all things); *teko porä* (Guaraní, good being); *ikigai* (Japanese, life purpose); *feta kgomo* (Sesotho, prioritizing the protection of life before the protection of wealth); *talkoo* (Finnish, collective work for a common good); *kanyini* (Yankunytjatjara, unconditional love with responsibility); and *minobimaatisiiwin* (Anishinaabeg, honoring all involved in creation and life). All our major human religions are infused with values of stewardship, reciprocity and the protection of life.[8]

These philosophies are broad and complex in their own right. But taken together, they represent visions of care and sublimity, aspirations to a world that venerates life and seeks to improve its quality. Such ideals can orient the development of alternatives and futures for society.[9]

Solutions in a complex world

The illusion of mastery and the logic of control extend from nature to society. Blueprints, manifestos, policies and plans carry the dream of surgical precision. Yet we cannot simply modify the world according to our wishes. There is no formula for the future.

Embracing complexity means acknowledging the limits of our ability to shape our societies, and looking at the political world in a different way. In an interconnected world, no action is isolated or neutral. Every decision will ripple across systems, with unanticipated and unpredictable effects. All actions shape and co-create the world around us. Details gain importance. Every context is unique, infused with history and culture. Changes are not smooth transitions, but fitful and unpredictable.

Unless we pay attention to the intricacy of things, we can misread structural causes as symptoms and confuse noble impulses with solutions. To ensure we maintain our critical gaze, we can orient ourselves with some core principles. All transformation is a creative, iterative process, devoid of fail-safe rules.

Connection

The injustices of the world are vast, but there is one major reassurance: the solutions to many of our problems can be found in the same place. Efforts to tackle and adapt to climate change are opportunities to advance on multiple fronts: to confront poverty, erode inequities, reduce violence, improve public health, ensure access to dignified work, extend land rights, transform education, deliver justice to victims, heal societal rifts and strengthen communities. Climate violence can serve as a canopy for broader conversations about well-being, in all its dimensions.

Connection makes both truthful and tactical sense, given our intersecting crises have multiple sources and no single resolution. Ecology is in some way the medicine of systems, working not to address problems in isolation, but crafting alternatives that pay off in more ways than one. By transforming food systems, for example, we can improve nutrition, absorb carbon, ensure future water access and strengthen rural communities. By transforming our energy systems, we can extend opportunities for secure employment, tackle fuel poverty, reduce air pollution and slash emissions.

Collectively, we can assemble integral, imaginative and viable people's solutions that improve the health of all spheres of human life, redistributing wealth, time and power.

Humility

Our obstinate obsession with simplicity can give rise to reductive ways of thinking that see a world of clear causes, culprits and cures. When thinking about social change, we should approach the challenges that face us in a spirit of humility. There are no quick fixes to our mutilated memory, no silver bullets that will bury defeat, no neat exit routes from our ecological calamity. Neither are there smooth landings, free of any turbulence, in chaotic systems. Throughout history, social transformation, whether from above or below, has always been convulsive and unpredictable.

Changing the world is fundamentally challenging. It is easy to get things wrong, or to lose sight of our honorable intentions that may return as unwelcome implications. The success of any social initiative depends on multiple factors, particular to each context. The same policy on infant nutrition can find success in Tamil Nadu and failure across the border in Bangladesh.[10]

Although there may be some win-win options, all major trans-formations carry dilemmas, trade-offs and difficulties. Sustainabil-ity problems, rather than puzzles to be solved, are challenges to be negotiated, riddled with knotty questions. What is the balance be-tween efficacy and ethics? How do we transform the energy system without extending energy poverty? How do we equitably down-scale major industries, such as the oil industry? How do we trans-form the food system without hiking prices for the poorest families, or loading even larger burdens on agrarian communities? What do we do with key industries (cement, steel and aviation) where low-carbon solutions are hard to find?

Simplistic thinking tempts us into the binaries of "good" and "bad" ideas, sustainable or unsustainable choices. But no proposals are free from errors or imperfections. Every promising idea, whether it be a basic income scheme or a carbon tax, carries potential to im-prove things, but also to reinforce inequities. If implemented with-out careful thought, consultation and participation, great ideas can fall flat or bear counter-productive consequences.

Without appropriate application, eliminating fossil-fuel sub-sidies can result in soaring energy bills for the poorest sectors of the population. Equally, improvements in energy efficiency, designed to save energy and lower emissions, can cause a rebound effect. This occurs when efficiency is accompanied by increasing energy use. Imagine a small firm reduces its bills by $50,000 through saving electricity; it can then use that money to buy more things (flights, equipment), generating more emissions.

Under our dominant economic model, declines in economic growth are associated with highly negative social effects. Reducing people's working week, unless complemented with other policies, can increase the available time for wealthy families to take high-carbon holidays. In addition, increased leisure time, without poli-cies to challenge gendered expectations, could lead to even heavier burdens on women.

Many proposed ecological solutions generate new problems. Dams, seen as regulators of water levels, accumulate destructive impacts upstream and downstream. Synthetic fertilizer, the antidote to decreasing fertility, degrades soils in the long term. Nuclear power, promoted as an endless source of mass power, carries with it endless risks.

Certain solutions to the ozone crisis proved, in retrospect, to be problematic from a climatic perspective, as they contributed to global warming. Certain energy-efficient measures introduced in buildings can prevent air circulation, causing health risks through the buildup of pollutants.[11]

Nor are environmentalist impulses necessarily socially just. For example, in a world where construction and engineering industries are disproportionately managed by men, green jobs without gender justice may mean "jobs for men."

Green transitions have often been accompanied by unjust consequences. Cities have climbed sustainability indexes while also ascending poverty and homelessness rankings.[12] Many of the policies that have made cities greener have also made them more unequal.[13]

Coal-plant closures, touted for their environmental benefits, have often been made without compensatory measures for workers or communities, exacerbating energy poverty, unemployment and already precarious living conditions.[14]

Ecological answers are not necessarily intuitive. Over the last centuries, guided by the impulse of control, many countries have implemented fire-suppression regimes to prevent forest fires. But suppressing fires leaves forests with major fuel loads, making them incredibly flammable. Today, many ecosystems have a widespread deficit of fire, and we are increasingly acknowledging that periodic fires are vital for forest health and biodiversity.[15]

One of the prime lessons of ecology is that we should be attentive to all relationships and dependencies, no matter how small or invisible. Future technologies, from smartphones to computer chips, all have material impacts and interactions, requiring the extraction of lithium, heavy rare earths, tantalum and copper. The energy required to maintain internet servers and digital devices is enormous; within eight years, the communications industry could account for a fifth of global electricity consumption. Touted solutions such as self-driving cars require huge amounts of computing power; the control systems of a single autonomous vehicle can consume electricity amounts equivalent to having up to 100 laptops running continuously.[16]

All these reminders should not turn us towards the paralysis of guilt, or force us to aspire to purity. We will never be able to transcend the messy complicities that complex systems imply. Instead, we should aspire to be attentively enthusiastic, doing our best to make contributions, not technocratic impositions.

Radicalism not romanticism

As we work to articulate alternatives, we should recall that every concept can be appropriated, every ideal twisted. Even seemingly benign ideas such as positive thinking, resilience and happiness have been used to naturalize injustice and justify inequality.[17]

Romanticizing any idea, initiative or state is futile. There are no flawless solutions, just worthwhile initiatives by imperfect human beings. We should draw critical inspiration from diverse models, while admitting that every attempt has rough edges, deficiencies and difficulties.

Bhutan's vaunted economic proposals coexist with severe discrimination against Nepali minorities: ecological arguments have been used, portraying the Bhutanese as a "species" endangered by Nepali culture and its inherent nature-destroying practices. The Masdar project in Abu Dhabi, promoted as the world's first zero-carbon city, was built by migrant laborers subjected to inhumane working conditions. Scandinavia's much-praised societies offer egalitarian inspiration, but Norway's reliance on oil revenue, Sweden's major arms-manufacturing industry and the region's institutionalized exclusion of foreigners should raise questions. Costa Rica's leadership on environmental issues has been accompanied by severe water pollution, waste injustice in slums like La Caprio and the unjust impacts of hydroelectric projects.

Our enthusiasm should be qualified with realism. We should

be wary of seductive panaceas and simplified responses, rejecting approaches that sell the causes of crisis as their correctives. Green growth, green capitalism, climate-smart agriculture, smart cities, post-carbon urbanization, the circular economy and ethical consumerism: all of these carry worthwhile elements, but can often be extractive practices with different labels.[18]

Many of the conversations around preparing for climate change, for example, revolve around building infrastructure: wider storm drains, higher seawalls, stronger levees, levelled-up beaches, fortified dams, larger desalination plants and robust electricity networks. Although these can be valuable buffers, fixing our focus on engineering can turn our attention away from the root causes of people's vulnerability.[19]

The sense of security associated with buildings can often hide the fact that infrastructure is not impenetrable. Seawalls crumble, dams are breached.[20] Some of the worst tragedies in history have occurred when climate extremes have collided with infrastructural failings. In 1975, dam failures in the wake of a typhoon killed over 230,000 people in central China. In 2006, over 100 people in the Indian city of Surat were killed when dam operators failed to reduce reservoir levels in the run-up to monsoonal rains. In Bangladesh, one adaptation project caused major displacement, as inappropriately installed polders and dykes neglected monsoon rainfall, waterlogging land.[21] Intrusive engineering measures can exacerbate erosion, damage riverine ecosystems and load vulnerabilities on communities.[22]

Infrastructure is also tremendously expensive, resource-intensive and notoriously hard to implement. After experiencing a tremendous flood in 1966, Venice took over 50 years to design, approve, fund and start building an ambitious water-protection plan.[23] These delays are not unimportant, particularly when climate change's inherent escalation of threats very quickly renders infrastructure outdated. Architectural designs are informed by frequently obsolete predictions. Previous estimates saw around 16 centimeters of sea level rise as the most probable scenario for the end of the century. Today, many scientists project likely rates reaching and perhaps exceeding as much as two meters by 2100.[24] Under such fluctuating conditions, we simply cannot engineer our way out of every problem, armouring the entire planet.

Technology is another area prone to over-simplistic thinking. Although technologies offer potential, their aura of efficiency and modernity can breed excessive optimism and reassurance, as we feel we can innovate our way out of catastrophe. We must be careful. The historian Malvin Kranzberg outlined one of the principal laws of technology: "Technology is neither good nor bad; nor is it neutral." The benefits of technology rely on their implementation, context, nature and ownership. In agriculture, technology can mean participatory breeding and seed-sharing between farmers; but it can also mean synthetic biology, transgenics and industrial hybrids. All technologies require relentless oversight to ensure they align with human purpose.

We need to be radically realistic and realistically radical. As Angela Davis explains, radical means "grasping things at the root." The writer Wendell Berry argued that "being radical enough" was synonymous with "being thorough enough."[25]

Avoiding false solutions

When it comes to climate violence, we need to be particularly alert to false solutions. These represent the fool's gold of climate change: diversionary proposals which produce new hazards, rather than tackling the root causes of the crisis. False solutions are often manifestations of magical thinking, led by a faith in the salvation offered by human ingenuity or the power of markets.

They include adaptation proposals such as the creation of floating island nations, or the production of infinite water through desalination. Perhaps the most salient "false solution" is biofuels. In a world of intolerable hunger and rampant deforestation, many biofuels are derived from foodstuffs (maize, sugarcane, oils), grown through the clearing of forests. From an energy perspective, the production of biofuels involves using energy to grow food, then applying energy to convert that food into fuel, then using energy to burn that fuel for energy.[26] Yet in spite of such illogicality, countries and companies are planning to expand supplies of "green biofuels," acquiring swathes of land to establish plantations.[27]

But another arena of proposals offers risks perhaps even greater than biofuels. Pessimism about our ability to bring down energy emissions, coupled with a reluctance to tackle what drives them, has drawn policymakers to carbon-dioxide removal and geoengineering.

Carbon removal

Carbon removal (also known as sequestration, or carbon capture and storage, CCS) is the process of removing carbon from the atmosphere to prevent it from trapping heat. Proposed versions of carbon removal include direct air-capture technologies that suck carbon from the air.

The most touted type of CCS is a technique known as BECCS: bio-energy, carbon capture and storage. BECCS involves planting and harvesting copious amounts of plant mass (such as trees or tall grasses), burning that biomass in power plants to convert it into fuel and energy, capturing the carbon released in the process and then storing the carbon underground for millennia. Deploying BECCS at the scale necessary to tackle climate change would require around 500 million hectares of land, around a third of the world's arable land and an area larger than one-and-a-half times the size of India.[28] Copious amounts of energy, fertilizer and water would also be required to grow the necessary biomass. In climate scenarios where BECCS is rapidly adopted, we would need to use swathes of land the size of subcontinents, and levels of water equivalent to all the water used for agriculture.[29] Even assuming this were possible, there are significant indications that extensive BECCS would pose severe threats to food security, the safety and land rights of people, biodiversity conservation, soil quality and water availability.[30]

The scale is staggering. Staying below 2°C may require the establishment of a carbon capture and storage industry four times the size of the current oil industry within the next three decades.[31] Today only 17 CCS plants are currently operating, with the capacity to absorb a few million tonnes of CO_2 each year.[32] According to some scenarios, we may need to store 40 billion tonnes of carbon annually within decades. At such rates, we would have to build 250 new CCS plants every year for 70 years.[33]

Basic chemistry is relevant here. Fossil-fuel burning changes the nature of fossil fuels: the mass of carbon dioxide in the atmosphere is essentially three times the mass of its fossil-fuel source. For BECCS to work, we would have to find huge amounts of underground space to store carbon: around 60 billion cubic meters every year.[34] Given the stunted growth of CCS over the last decades, and the significant financial obstacles involved in its development, such an extraordinary boom seems highly unlikely.

Yet enshrining BECCS as a future solution is both an instrument of procrastination and an additional way of transferring responsibility to future generations. It forces those unborn, who are already bound to live in comparative ecological poverty, to foot an impossible bill, which could reach $535 trillion.[35]

Geoengineering

Geoengineering would entail technologically changing the climate: the most ambitious engineering project in human history. There are various ideas, including fertilizing oceans, deploying cloud-forming ships and issuing weather-enhancing balloons. The principal idea is, however, called albedo modification, or solar radiation management. This would represent an attempt to increase the reflectivity of the Earth through pumping sulphur aerosols into the atmosphere to block the sun's rays.

Geoengineering is perhaps the epitome of the dream of control over nature: human orchestration of the rain and the sun.[36] The word "engineering" may suggest precise control and predictability, but all these proposals are entirely experimental. We have never toyed with our planet's climate in this way, and there are strong reasons to suspect that geoengineering techniques could risk unimaginable damage to the web of life.

Furthermore, if we were to take up albedo management, we would have to do it every year, forever. As physicist Raymond Pierre-humbert observes:

> Any kind of geoengineering of the Earth's reflectivity...is not a one-time, one-off event, but something which would have to be repeated in perpetuity.... What is the morality of committing 10,000 years of future humanity to maintaining an activity year in and year out without fail? What is our track record as a species of maintaining any technological activity for more than a century or two? [37]

Such technologies would lock us into millennia of vigilance. What if we rely on it and it fails? Geoengineering could be tremendously high-risk, opening opportunities for abuse, profiteering and weaponization.[38]

Yet despite nonexistent debate about these false solutions and their implications, mainstream climate models are actually reliant on their success. The majority of governments endorse emissions trajectories entirely dependent on the widespread uptake of carbon-dioxide removal and negative emissions technology. Various trials have already been scheduled for 2018.[39]

Such gambles may seem far-fetched today. But, as climate violence intensifies, will our calculation of their morality change? Will ecological desperation make us more amenable to extreme intervention?

But there may be even more bitter realities. The rising interest in geoengineering is also a reflection of our failure to curb emissions over recent decades. The models have made it into a reality. As fantastical and dangerous as negative emissions technologies may be, they may be rendered indispensable if the status quo remains unchecked.

Democracy, diversity and accountability

The mentalities of the past have handed us a legacy of uniformity, a firm belief in the possibility of implementing replicable templates in every context. This closed-mindedness has neglected the pluralism of human society and the inapplicability of one-size-fits-all approaches.

Ecological systems thrive through diversity. Where diversity flourishes, risks subside, whereas systems that lack it are highly vulnerable to disease and danger. Diversity is coded within us. The process of meiosis adds genetic variation to our cells, as chromosomes cross over and recombine. All of us are composites, impossibly elaborate embroideries of lineages.[40] In ecology, the ecotone refers to a place where biomes meet, such as marshlands, mangroves and estuaries. These areas, where borders blend, are filled with extraordinary richness of species.[41]

Our solutions should be as plural as the world we inhabit and the world we desire. As Sunita Narain notes, "It is only when...biodiversity is lived that it will live."[42]

Acknowledging this diversity means ensuring that the perspectives, visions, needs and values of all must be given weight. Those closest to the land should be closest to decisions regarding that land.[43] This leads us to the principle of accountability. If we want to decipher the justice of a decision, we have an obligation to listen first to the voices of those most affected by that decision.

Our moral judgments flow from the voices we heed. In late 2015, diplomats negotiated for weeks in the outskirts of Paris. On 12 December, the Paris Agreement was adopted by representatives of 196 states. United Nations Secretary General Ban Ki-Moon called it a "monumental success for the planet and its people." French president François Hollande described it as "a major leap for mankind."

Simultaneously, Nigerian activist Nnimmo Bassey defined it as: "[a betrayal] of the poor, the vulnerable and all those already suffering the impacts of climate change. [Paris sets] the stage for a climate-changed world, and [does] little about averting it."[44] Ethiopian activist Azeb Girmai rued "the saddest day for all the poor people in the world facing loss and damage day-in and day-out."[45]

Who is right and wrong here? The difference between these positions is one of ethical grounding. All ecological arguments and judgments reflect particular moral positions.

Yet often we ignore this, speaking about ecology in ways that suggest an abundance of singular technical and scientific solutions.[46] This depoliticized ecology frees us from choice, but a more honest ecology might show us the value-laden choices we will have to make.

Is food a right or a commodity? What sacrifices are necessary? What is politically prudent, or expedient? Do we see our territory as a terrain to be exploited continuously, or as a space fundamental to long-term survival? What are the sources of well-being? Do we prioritize bottom-up or top-bottom approaches? How much should we embrace technology? How do we regard people?—Do we see populations as things to be governed, people to be improved, rational actors to be guided or humans to be emancipated?[47]

All major issues—whether it be urban planning, healthcare provision or the introduction of technology—involve weighing up visions. We have to work collectively, guided by ecological literacy, to make tricky decisions.

All solutions will reflect the interests of their origin, and so our stances should be grounded in the urgency of the most afflicted. The Argentine activist Juan Carr explains:

> I know that pain is in charge, that they who suffer know. I know that approaching a person that suffers is like entering a temple. I know that pain dislocates and robs a person of the coldness necessary for calculation. But the person who suffers knows more than me, more than everyone.

We need our ecological visions to reflect the priorities and vulnerabilities of those most affected by climate violence. We need solutions where justice, and those most in need of it, is not an afterthought—where those rendered most vulnerable by our economic system are prioritized so that they will be the first to benefit from the transition to an ecological economy.[48]

History shows us that power structures are stubbornly enduring. The architecture of oppression is resistant and hard to dismantle. Unless we prioritize justice, all new proposals are likely to entrench parts of the status quo. To commence this shift, we will need to reverse the exclusion of marginalized voices, which has skewed mainstream visions and impoverished our ability to understand the world.

As activist Erik Leipoldt reflects:

> The perspectives and experience of life lived with disability must be part of the sustainability debate if we want effective action...the disability experience of heightened vulnerability and dependence is a useful magnification of the human condition...[and] a model for a sustainable world...we represent life as it really is: uncertain, imperfect, and worth living...a society, through free interaction with people with disabilities in all its areas of life, would be more mindful of its essential interactions with and dependence on others and on the environment. A sense of individual separateness would diminish.[49]

Leipoldt's words ring true. Ultimately, diverse perspectives enrich our ideas and endeavors. Inclusive approaches are more open to learning, acknowledging mistakes and allowing modifications. Community-based early-warning systems, participatory planning to ensure disaster resilience or locally led efforts to protect common resources: all these are typically far more successful than approaches that involve centralizing power.[50]

Sustainable solutions fail dramatically when installed without local leadership or attentiveness to the particularities of each population. Houses are built with inappropriate materials that don't dissipate heat; solar ovens installed to reduce dirty energy use are disavowed when their size prevents social cooking.[51] In late 2016, the Peruvian region of Rosario was experiencing an intense drought. High temperatures, combined with a delayed rainy season, were killing livestock. Local villagers began to grow suspicious of the expensive early-warning systems nearby, installed to warn residents of dam ruptures upstream. Community members associated the lack of rain with the machines, and ended up destroying them. Three days later, the rains returned.[52]

The paradox of pace

The mathematics of carbon demand urgent emissions cuts. We do not have centuries to make the transitions we need. Climate change throws us a ticking bomb. But while time is running out to act, transformations of the magnitude required take time. The construction of new infrastructures, cultural changes and behavioral switches can take decades.

The playwright Vaclav Havel, a key figure in Czechoslovakian resistance to communism, later reflected:

> I realize with fright that my impatience for the re-establishment of democracy had something almost communist [or rationalist] in it.... I had wanted to make history move ahead in the same way that a child pulls on a plant to make it grow more quickly. I believe we must learn to wait as we learn to create. We have to patiently sow the seeds, assiduously water the earth where they are sown and give the plants the time that is their own. One cannot fool a plant any more than one can fool history.[53]

Havel's words hint at the different speeds of the world. There are differences between the pace of stars, the pace of mountains, the pace of trees, the pace of buildings and the pace of humans. The cycles of life do not follow the cycles of politics and business. Quarterly profits, the four years of an electoral mandate, or the urgencies of a news day, are largely incongruent with the time-scales of the planetary challenge.[54]

We drastically need more coherent political horizons. With climate violence, gradualism is implausible as a strategy. Patient successes equate to defeats.[55] Yet a frantic, save-the-world mentality can bring with it a worrying indifference to process. As we have seen, our climate crisis involves the combination of both atmospheric and social conditions. Unless we accompany attempts to pull down emissions with efforts to erode injustice, we will fail on both counts.

The business of boldness

The stark mathematics of carbon budgets, and the vast inequities across our planet, compel us to be bold. We need proposals and approaches that are designed to yield results before impacts overtake us. Piecemeal policies will not end poverty, destructive development or our carbon crisis: only addressing the ideas and structures driving them will do that.

We must work to put forward actions proportionate with the scale of our crisis, crafting radically new models of production, consumption, development and urbanism. Climate change entails transformation and solidarity on a scale that has no precedent.

Yet even this is possible. In the Indian state of Uttar Pradesh, 800,000 people planted 50 million trees on a single day in July 2016.[56] Many have urged for such large-scale solutions: an ecological equivalent of the Manhattan Project, a Marshall Plan for the climate, an Apollo programme for renewable energy. In a sense, they are right. What these examples indicate is that, under enough pressure, entire economies can be dramatically and speedily shifted.

During the Space Race, 400,000 people worked at NASA, receiving four percent of the US federal budget. During World War Two, to combat the fascist threat, entire economies were reconfigured. In the United States, gardens and empty lots were turned into farms; these "victory gardens" provided 40 percent of national food. Industries were nationalized, as car manufacturers were forced to produce military equipment. From toothpaste to cooking grease, all was recycled. Lessing Rosenwald, chief of the Bureau of Industrial Conservation, called on Americans "to change from an economy of waste...to an economy of conservation." Thousands of schoolchildren, youth groups, trade unions, neighborhoods, joined campaigns to collect scrap metal and maintain gardens. Automobile factories were converted to produce tanks and bombers. With rubber and gasoline rationed, people took up cycling and drivers were encouraged to participate in carpools. Those driving alone recreationally were fined. Political parties even ordered their candidates to support the cause by hitchhiking to campaign rallies.[57]

Unprecedented technological changes, when accompanied by the right social and political shifts, have also taken place before. In the late 19th century, the world's industrial cities were defined by horse-drawn carriages. In 1894, *The Times* warned: "in 50 years, the streets of London will be buried under horse manure." Within a few years, horse-drawn carriages were in a minority. In 1980, consultancy firm McKinsey was commissioned to estimate the future mobile-phone market for the year 2000. The firm projected 900,000 mobile-phone users; it ended up being 109 million.[58]

Although there are significant differences between these eras and today, these examples show us both the malleability of society and the uncertainty that surrounds what is possibile.

CHAPTER 13

A MOSAIC OF ALTERNATIVES

What we do to the land, we do to ourselves.
WENDELL BERRY

▪ ▪ ▪ ▪ ▪ ▪ ▪ ▪ ▪ ▪

Decolonization is not a category...a property, or a thing in itself—
it is a queering of things, of relationships, of assumptions. A stunning
"what if"...an ethical call.... It is the rift in the spaces of power
suggesting that there are many other ways to become with the world
that disturb the hegemony of humans and human interests.
BAYO AKOMOLAFE[1]

▪ ▪ ▪ ▪ ▪ ▪ ▪ ▪ ▪ ▪

Healing country heals ourselves,
and healing ourselves heals country.
JUDY ATKINSON

▪ ▪ ▪ ▪ ▪ ▪ ▪ ▪ ▪ ▪

WHAT FOLLOWS IS NOT an exact formula or a plotted route to climate safety. It is a brief map of 12 terrains of possibility, which may represent existing and potential pathways for a transition.

An economy of life

The dominant story of economics describes a world of perfect rationalities and obtainable equilibria. Financial value is its mast: a neat, singular compass for all decision-making. The recipes for prosperity are reassuringly simple: growth, expansion, profit maximization, competition.

205

The damage done to the sources of life demands a bold challenge to this story, which has wrought a destructive development model, restricted the vibrant possibility of human society and closed room for other forms of economic relations.[2] We need a bold shift in perspective, one that tames the thirst for growth and understands forms of prosperity beyond it.

To begin this transformation, we must first change what we measure and strive towards, asking ourselves: what do we deem valuable and invaluable? The economist Ana Esther Ceceña writes:

> One of the separations implanted by capitalism and Western thought is the one between the production and reproduction of life. We are accustomed to...sacrificing everything for production...[to] guaranteeing economic indicators...but we don't talk about living better. In many cases people are asking and asking governments: if the economy is going so well, why are we doing so badly?[3]

Profit, size and growth are very limited filters for viewing prosperity. We need to reassemble our compass and apply more ecological thinking to the economy. Our task is to replace the love of size with a love of sufficiency; the love of more with the love of better. To value qualities over quantities. To improve instead of maximizing.

We must go beyond GDP to embrace a plurality of values that assess our individual, collective and ecological well-being. Economic decisions should be made with firm consideration of our material footprints, the health of our ecosystems, the equality of our societies and the equity of our relationships.[4]

Deprivation should be understood not as a singular absence of income but as a composite of multiple unmet needs: subsistence, protection, affection, participation, education, identity, time, leisure, solace, freedom and creation.[5] Instead of striving to keep populations above outrageously low poverty lines, we can follow the lead of diverse social organizations that have proposed a line of dignity, a standard which incorporates cultural, political and environmental rights.[6]

The economist Manfred Max-Neef offers a useful set of values for rethinking our economy:

1. The economy is to serve the people and not the people to serve the economy.
2. Development is about people and not about objects.
3. Growth is not the same as development, and development does not necessarily require growth.
4. No economy is possible in the absence of ecosystem services.
5. The economy is a subsystem of a larger finite system, the biosphere, hence permanent growth is impossible.

These five postulates are underlined by one overarching principle: "No economic interest, under any circumstance, can be above the reverence for life."[7]

Under such a view, the fundamental economic challenge shifts from how do we grow more to how can we live well within limits?[8] How do we ensure the basic needs of over seven billion people are met, without running roughshod over the planet's boundaries? How do we improve well-being while simultaneously eliminating emissions?

We should strive toward what economist Kate Raworth calls the doughnut: a space for human flourishing where planetary limits are the ceiling and basic human needs are the floor.[9] How can this be achieved? Primarily, we need to throw into question the uniform model of development: industrialize, modernize and grow. Instead we need a bloom of endeavors: restructuring the economy, reducing needless consumption, redistributing resources, restoring ecosystems, regulating production processes and rewriting new relations with the life systems around us. We need to begin identifying and confronting the roots of suffering—unequal trade, debt burdens, colonial legacies, labor exploitation—that underlie the contemporary system.

Through such efforts, we can weave together economies that are attentive to what matters and not blind to what injures. Economies of life that leave smaller ecological footprints and more expansive "civilizing footprints: imprints of care, creativity, and culture."[10]

The supposed tensions between conserving jobs and conserving the environment should be driven to absurdity. We have every chance of radically transforming the world of work, creating decent jobs that confront poverty and secure livelihoods that protect ecosystems rather than plundering them. Rapidly shifting to the zero-carbon society we need to address climate change is an enormous task that will require full employment.[11]

We could strengthen local economies and generative businesses, providing jobs that nourish people and bring real value to communities: agroecological farming, ecosystem restoration, renewable energy development, efficient resource use, public transport, small-scale manufacturing, arts, care, education and recycling.[12]

We could tackle the violence woven across the world of work. Today there are more humans in forced labor than were transported by the transatlantic slave trade. Annual workplace "accidents" cause more deaths than warfare.[13]

There are plenty of innovative business models to experiment with and abundant examples to draw inspiration from. The world over, communities are crafting economic alternatives relating to education, care, gastronomy and ecological restoration.

In the Mexican state of Puebla, the cooperative Tosepan Titataniske brings together organic coffee farms, cosmetics workshops, ecological tourism initiatives, ecological construction firms and a cooperative bank with over 30,000 members. The mining company Umicore has moved from standard extraction to the recycling of e-waste. Initiatives like Cargonomia in Hungary, connecting together an organic farm, a biking cooperative and a courier association so as to deliver organic produce to city residents involved in an association to promote rural agriculture. In Japan, the mutual aid network Fureai Kippu provides care to elderly populations through a time banking system.[14]

From an ecological perspective, we need to reshape our profligate economic metabolism: its consumption of resources and energy, and its return of waste. This entails major efforts of investment and disinvestment, of growth and contraction: expanding the parts of the economy we value, diluting those we don't. We can produce less and design better. Instead of internalizing externalities, we can support other modes of production that don't produce them.[15]

We could strive for a world of equal opportunities and conditions, where people have the freedom to pursue diverse ways of being and of contributing to human society. Universal job guarantees, a global minimum wage, global basic incomes, global guaranteed services: all of these policies have potential.

We could shift public money away from polluting to life-sustaining industries. Today, hundreds of billions of dollars in public funds flow annually toward the fossil-fuel industries. In Brazil and Indonesia, subsidies for industrial agriculture are 100 times greater than money disbursed for forest protection.[16] In Australia, mining companies receive in fuel-tax credits double the public spending afforded to environmental and biodiversity programmes.[17] Resources could be transferred so as to transform our energy system, support the restoration of ecosystems and decarbonize transport.

We can democratize financial institutions, encourage cooperative banks and transform the money-creation system.[18] Today, most of the money in the economy is created by banks. When banks make loans, they create money as debt. These loans largely go into property and finance. Rather than private money creation, public money creation could be directed toward education and social care.

We can rearrange our taxation system, which is tilted at the moment toward taxing labor and benefiting the wealthy. A more ecological tax regime could reposition taxes away from people's work toward resource use. Taxes on carbon; levies on fuels; charges on fossil-fuel extraction; taxes on robots; land-value taxes; higher sales taxes for products with lower ecological standards: all of these could generate revenue. In countries highly reliant on mining commodities or fossil fuels, these industries could be gradually phased out, with increased royalty arrangements financing the development of economic alternatives.[19]

We could tax wealth better, closing loopholes and offshore havens. Simple reforms can yield significant results. Studies have shown that simply ending tax incentives for transnational companies in Sierra Leone could increase the country's education budget sevenfold.[20]

All these are, of course, just ideas, and realizing them would be no easy task. Nor is their application universal. Some people live in the extravagance of overconsumption, while many live in the indignity of underconsumption. The poorest are entitled to consume more, and to ensure climate safety, the highest-consuming nations need to downscale their consumption to sustainable levels, living better with less.[21]

We are bound to be averse to giving up on the aspirations of boundless consumption, innovation and riches, for the narrow story of prosperity promised lifestyles of abundance. Consumerism and material wealth occupy almost mythical roles in our culture. But, as psychologist Tim Kasser argues, we can replace the "goods life" with the "good life."[22] Research across medicine, psychology and happiness economics has found converging evidence over what leads to human well-being and longevity: strong social connections, feeling of purpose and healthy environments.[23]

The challenges are undeniable. Currently, not a single country meets its citizens' basic needs while maintaining sustainable levels of resource use.[24] But we must begin reconfiguring institutions and experimenting with different approaches, policies and arrangements. Countries such as Costa Rica and Cuba have outperformed much richer nations on indicators such as infant mortality and life expectancy as a result of their social policies.

Justice

When it comes to ecology, impunity has imposed itself over justice as the default outcome. In such a context, environmental crime is more profitable than change. Wrongdoing is routine: from the ships of poison that dump waste illegally on the poorest communities to the automobile manufacturers that have lied about emissions standards; from the fossil-fuel companies who publicly discredit climate science while internally acknowledging its truth to the countless murders of environmental defenders.

The absence of justice, an unwritten licence to pollute, costs lives. In 2015, mass fires incinerated over two million hectares across Indonesia. The fires, generated by illegal burning predominantly driven by agricultural and commodity expansion, fostered a catastrophic haze which killed up to 100,000 people prematurely in Indonesia, Malaysia and Singapore.[25] All charges were eventually dropped against 15 companies connected to the fires.[26] One study found that excess emissions from Volkswagen's cars led to 1,200 early deaths in Europe, over a period of seven years.[27]

When justice is applied, rarely is it extensive enough to prompt behavioral change. In 2009, Shell agreed to an out-of-court settlement with reparations payments of $15.5 million for communities in the Niger Delta; this represented just four hours of the company's profits. Exxon Mobil violated the US Clean Air Act 16,386 times over an eight-year period; they received a mere $21-million fine.[28]

We must work to bring impunity to a close, ensuring that those who have contributed most to the ecological crisis must contribute most to its rectification. States must reinforce corporate accountability, ending the ample protections that exempt transnational corporations from justice. Extended producer responsibility laws, which deem companies responsible for their products throughout their life cycles, should be adopted. Constitutions should guarantee the rights of nature, and the environmental rights of citizens, including the rights to clean air, to safe environments and to participate in dialogue over the environment. Regulations around environmental limits and resource use should be introduced and bolstered.

Instead of permitting companies to pay for compensation after pollution, we need bold legislative efforts to prevent the insidious crime of contamination. Production processes have normalized the exposure of human beings to hazardous chemicals and heavy metals. The memory of pollution is long. Once toxins are made, they easily disperse, flowing into rivers, seeping into sediments, dissolving into air, entering the circulatory system of the planet. They enter our lungs, our air, our food and our water. These toxic materials bioaccumulate and biopersist, posing carcinogenic, mutagenic and reproductive risks.[29]

All our bodies carry toxic burdens, transposing them to new life. Babies are born into a polluted world, with polluted bodies. Breastmilk has been found to have alarming levels of toxicity. One study of umbilical cords detected an average of 287 industrial and agricultural chemicals.[30]

Virtually no reliable safety data exists around most of the thousands of synthetic chemicals to which we are exposed. As doctor Hireesh Chandra argued in the aftermath of the Bhopal disaster, corporations should not "be permitted to make poison for which there is no antidote."[31] We need serious measures to reduce the toxification of our bodies, landscapes and waters.

Environmental justice is not a punitive call for the incarceration of a group of clearly defined environmental criminals. It is a call for a different response to violence that departs from our penal and castigatory status quo. Justice should be transformative, a social process that strives to clarify the context of wrongs, build plausible repair, help heal those afflicted, ensure accountability and transform the conditions that allowed injustice to take place. All collective injuries require collective treatments.[32]

Nourishment

Food is the basis of nourishment and life. But much of what we currently eat comes from a system of intensive industrial agriculture, oriented to produce financial fortune before food. In this model, nature is a productive machine, where farms serve as factories, lands as grown assembly lines, soils as mines of fertility and rivers as infinite pipelines.

The measure of a system should be its effectiveness, but the industrial food system does not feed all humans. Billions of people lack adequate nutrition. Even those closest to food often lack it, for most of the world's hungry are farmers.[33]

For over a century, our issues related to food have been framed as problems of production alone. States have single-mindedly focused on increasing productivity. The much-touted Green Revolution used high-yield strains, artificial fertilizers and motor pumps to ramp up production. Although politicians lauded this model, its short-termist mindset left high costs in the form of exhausted soils, depleted groundwaters and parched lands. A third of global farmland has been destroyed through erosion and degradation over the past 40 years.[34] But these ecological maladies were masked through irrigation, animal feed imports and heavy doses of agrochemical inputs.[35]

The monolithic focus on productivity obscures a more holistic view of agriculture. Yields are just a part of the picture. The sustainability of agroecosystems, community health, farmer income, communal well-being, nutrition and animal welfare are all also important. When all these elements are included, ecological agriculture can match or surpass industrial yields.[36] One major study of hundreds of sustainability-enhancing agricultural projects over 57 countries found that, on average, crop yields grew by 79 percent.[37]

Industrialized agriculture does not even produce the majority of the world's food. Across the world, industrialized agriculture uses over three-quarters of arable land, to feed a mere 30 percent of the globe. The peasant food web—the small-scale fisherfolk, pastoralists and peasant farmers—uses less than a quarter of the land to feed the remaining 70 percent.[38] In Russia, rural cottages (*dachas*) produce two-fifths of the country's food, including two-thirds of its vegetables, half its milk and 80 percent of its fruit.[39]

Yet for all the successes of the peasant food web, the industrial food model is increasingly encroaching on remaining land, eroding rural communities. The money associated with food flows to those beyond the harvest, siphoned off by food processors and retailers. These dynamics have meant that today, in many contexts, to be a small-scale farmer is to harvest debt and poverty. Across sub-Saharan Africa, three-quarters of all undernourished children live on smallholder farms.[40] In the state of Punjab, over four-fifths of agrarian families live in debt.[41] These farmers are burdened with the rising costs of herbicides, pesticides, machinery, fuel and seeds. Rendered unviable, smallholder farmers are disappearing, abandoning their communities to find a precarious living in cities.

For those farmers who remain, industrial agriculture is increasing burdens and restricting choices. Commercial pressures, debts and climatic shocks are contributing to alarming levels of mental illness in agricultural communities. In France, farmers experience suicide levels 22 percent higher than the average population. In India, 12,062 farmers committed suicide in 2015, according to state statistics; one every 41 minutes. Between 1995 and 2015, 318,528 farmers committed suicide. In Australia, a farmer commits suicide every four days. In the US, people working in agriculture, from fisheries to farms, experience higher rates of suicide than any other economic sector. The suicide rate in agriculture is five times higher than the general population, and double that of military veterans.[42] These rates are likely underestimates, for many suicides are unreported, or concealed as agrarian accidents.

The violence borne by farmers mirrors the treatment of the territories they tend to. Industrial agrarian models have enclosed lands, appropriated fishing waters, cleared native forests, drained wetlands, degraded soils and depleted previously fertile landscapes. Through practices of simplification, complex ecosystems have been displaced and replaced with monoculture plantations. Fields, stripped of trees and hedges, have been made amenable to machines. Livestock breeding has been transformed into intensive factory farming.

To grow, plants need macro- and micronutrients such as carbon, hydrogen, nitrogen, oxygen and phosphorus. Nitrogen and phosphorus in particular are the building blocks of life, shaping the fertility of soils. But through intensive industrial agriculture, their natural cycles have been drastically reconfigured. The intensive use of fertilizers, pesticides and herbicides has significantly deleterious effects on people and planet. Over half of the nitrogen applied is not taken up by crops, seeping excess nitrates and phosphate into water bodies and the atmosphere. Such processes are reshaping the chemistry of water bodies, overloading ecosystems, starving waters of oxygen and asphyxiating aquatic life. This agrochemical model is ultimately a self-defeating system. To compensate for soil degradation, we end up intensifying chemical inputs.

Industrial systems of agriculture have also dealt death blows to the web of life, eradicating animals and their habitats.[43] The diversity of our seeds, foods and nutritional traditions has plummeted. In 1903, over 300 varieties of corn were widely grown in the US; by 1983, the number had dropped to 13, and today the number is just 5.[44] In Kerala, the number of varieties of rice cultivated has diminished from 3,000 to 200.[45] Fundamentally, changing our food system means changing our paradigm. Agroecology contemplates more complex and diverse natural systems. Monocultures are replaced by polycultures that diversify resources, rotate crops and improve the coexistence of plants and animals. Farmers cannot simply produce commodities, but heterogeneous products. Ecological agriculture acknowledges the relationships that sustain ecosystems, and works to enhance these interactions. Soil, the skin of the Earth, is no longer seen as a factory of plants, but as a rich ecosystem, an encounter between our atmosphere, biosphere and lithosphere, which combine to give us sustenance.[46]

Agroecology uses more biodiverse, regenerative techniques of agriculture. Silvopasture, for example, integrates livestock breeding with agriculture by planting trees on pasturelands, which in turn fertilize the pasture. Agroforestry blends trees and crops in creative forest gardens. Such forms of land use curtail erosion, recharge groundwater, nourish soils, support biodiversity and absorb significant amounts of carbon.[47]

Land-management practices that replenish soils are prioritized, for a healthier soil retains more water, and is better at suppressing pests. Certain shrubs, trees and legumes, by fixing soil nitrogen, can maintain soil fertility. Flowers are planted to attract beneficial pollinators, and prevent pests. Cover crops, known as "green manure," are grown to return nutrients to the soil.

These more complex agroecological systems have also proven to be more resilient than their alternatives. Detailed studies into the impacts of major storms have shown how agroecology and polycropping practices can offer protection against the ravages of extreme weather. Ecological farms typically suffer far less damage than monocultures, and recover production far quicker. When Hurricane Ike swept over Cuba in 2008, diverse farms experienced losses across half of their farmlands, while monoculture plantations faced losses that were almost complete.[48] In times of drought, the healthier soils of agroecological farms are more adaptable, given their ability to retain moisture longer.

Supporting agroecology could also add a huge boost to carbon sinks. Rather than intensive plowing, which strips topsoil and releases carbon, we could prioritize no-till approaches. Since the dawn of agriculture, our soils have lost over 130 billion tonnes of carbon.[49] Farming can also provide an enormous service to carbon storage. Through techniques such as reduced tilling, compost application, enhanced grazing or cover cropping, farms can become carbon farms, and can significantly reduce methane emissions.[50]

Agroecology also demands a shift away from the production and consumption of resource-intensive foods. Nearly four-fifths of agricultural land is used to feed livestock, although they supply less than a fifth of global calories, and a third of global proteins.

Growing meat consumption is driving deforestation, water pollution and the collapse in biodiversity. For every million kilos of beef exported from South America, around 45,000 hectares of rainforest need to be cleared.[51] Such changes in land use, combined with the methane emissions of digesting ruminants, mean that livestock contribute to around 14 percent of emissions.

A healthier, more vegetarian global diet could deliver significant dents in food emissions, ecological destruction and global mortality.[52] IPCC research into land use has concluded that dietary changes are fundamental to staying behind temperature thresholds of 2°C.[53]

Agroecology also involves thinking about the wider spectrum of food, recognizing the injustice of unequal distribution. We grow enough food to feed at least 10 billion people. Huge amounts of food are squandered across the food chain, in fields, granaries, supermarkets, restaurants and homes. Around a third of seafood, a quarter of cereals, a fifth of all meat and around half of all vegetables and fruits are wasted.[54] If food waste was a state, it would be the world's third-largest emitter, after China and the US.[55] Major amounts of food, including a third of US corn, are converted to biofuels to drive cars.

Agroecological thinking requires us to heal the rift between rural and urban life, and to recognize agriculture as central for ecosystems, health, communities, culture and other beings. We need bolder visions that value agrarian work and encompass a composite improvement of life for farming areas.[56] Farmers need to be seen as core protagonists of society, rather than as disposable workers in a provisioning system for city residents.

As agricultural researcher Anil Bhattarai explains:

> Farming is becoming culturally unappealing, and materially challenging. Land has become opportunity for making quick money without cultivating the soil.... [For farmers] what matters is not only what happens in their fields, but what kinds of homes they live in, what kinds of education that their daughters and sons acquire, and what kind of healthcare they can access when in need.[57]

Today, the energy required for agriculture has been outsourced to machines, leaving farmers without lands, and lands without farmers. In industrialized countries, only one or two percent of the population works in agriculture. More sustainable models of agriculture are more labor-intensive. To ensure this transition, agricultural jobs need to be well-compensated jobs with improved conditions. But this is a different kind of agricultural labor: not the routine, menial work associated with industrial farms. In agroecology, farmers become educators, innovators and collaborators. Rather than chemical-intensive, agroecology is knowledge-intensive.[58] Extending rural democracy can allow for farmers to shape decisions relating to land-use strategies.

It also means building healthier rural communities. Today, many rural villages have been ravaged by the health impacts of agrochemicals, linked to augmented risks of sickness, cancer, reproductive health impacts and birth abnormalities.[59] Uncountable numbers of farmers have been lost to pesticide poisoning.[60] UN reports have linked over 200,000 annual deaths to acute intoxication from agrotoxins.[61]

By transforming the consumption and production of food, we can enhance nutrition, restore ecosystems, reduce emissions and extend life. We can build an agriculture that feeds rather than starves culture. We could encourage a fairer food culture where people have the right to healthy, equitable, sustainable and culturally appropriate food that does not harm them.

We can abandon a system of nutrient-poor, highly processed food, which is contributing to a major public-health crisis where diet-related diseases are on the rise. Obesity is less common in countries with strong, healthy food cultures and short supply chains.[62] We can shift a model of fast food, lonely cooking and discardable packaging toward another kind of eating culture: communal meals in the streets; slow food; communal bakeries, public kitchens and dining halls.

We can promote agrodiversity and reclaim our forgotten foods. Just 50 crops supply nine-tenths of the world's calories, fat and protein.[63] The average urban consumer will consume around 20 species of vegetable or fruit. Many rural family farms cultivate hundreds of species. In Mexico, researchers have documented over 700 species on some small family farms.[64] There are around 12,000 known varieties of potato, and 10,000 known varieties of rice.[65] Maintaining a wide variety of seeds allows farmers to blend yields and to adapt strands to particular conditions.

We can view food as a commons rather than a commodity. Agricultural engineer Jose Luis Vivero Pol has advocated the idea of "universal food coverage," where food access is conceived as being similar to education and healthcare. Like doctors or teachers, farmers could be employed directly by the state. Foodstuff speculation on international financial markets would be prohibited, supervised by human rights councils rather than trade bodies.[66] Everything is possible: it's just a matter of boldness.

The commons

For farmers to be able to grow sustainable food, they need land. But across the world, land ownership inequality is widespread, with most territory concentrated in the hands of the few. In the European Union, just 3 percent of farmers own over half of all farmland.[67] In Colombia, 1 percent of farms own 81 percent of the land.[68] Disparities in the control of land yield huge disproportions in the distribution of its benefits. In India, just 15 percent of landholders receive 90 percent of national agrarian income.[69]

How have we reached these levels of disparity? The conventional vision of land is rooted in old ideas of separation and individualism. We see land as property, handing owners exclusive rights of control. While there is an important role for private property, political systems and contemporary property law have wrought a world of dominion. Through processes of enclosure, what has been commonly shared has been appropriated into private hands. This process was designed to pull apart what is known as the commons. Commons are resources that we share, managed according to rules, norms and rights of use. Our air, our water, our lands, our forests, our atmosphere and outer space are all commons. Our knowledge, our artistic traditions, our historical memory, our human heritage are examples of more immaterial commons.[70]

For too long, economists and politicians assumed that the protection and stewardship of shared resources were wasteful, contravening the logic of efficiency through competition. But extensive research shows this archetype bears little relationship to reality.

Historically, humans have sustainably and productively managed lands, forests, pastures and rivers as commons. Today, people across the world manage a range of shared goods in convivial, creative and sustainable ways.

The commons hand us a prism through which to see the world, to prioritize access before property. Shared resources bring shared responsibilities and benefits. By valuing the commons, we can bring more democratic forms of control to bear on what is most valuable, from our neighborhoods, to our information, to our food.[71]

This change of paradigm, and its associated devolution of power, yields major ecological benefits. Through the commons, people see themselves as stewards of ecosystems, rather than as owners of land. In cases of climate violence, communal decision-making offers a buffer to manage difficult circumstances. In Andhra Pradesh, farmers' groups work collaboratively to measure soil moisture and take collective decisions on water collection. With winter storms in Mongolia's Gobi Desert gaining force and frequency, Uvurkhangai herders have formed communities to better manage pastures, pool resources and ease land degradation.[72]

We could encourage communal institutions and cooperative approaches across all spheres of society. Rather than private digital platforms that extract wealth from social interaction, we could have the same services with different business models. By conceiving some of the services we most require—care, food, information, urban space—as commons, we can begin to reclaim our common wealth.

Energy

The defining story of our ecological crisis involves energy. Energy concerns the work done by humans and nonhumans, the invisible labor of ecosystems and its members.[73] Over human history, societies have harvested energy in various forms, and applied it to develop their societies. Fossil fuels, and their energy surpluses accumulated over millions of years, formed the basis of our industrial society. We have built the present by burning copious energy deposits, overheating the atmosphere in the process.

This burning must rapidly be wound down to ensure any chance of climatic safety. We have to move from a dependence on nonrenewable fossil-fuel stocks to renewable solar and wind flows.

But the pace of the transition needs to be unprecedentedly rapid. Historically, energy transitions have taken centuries. Renewable energy will need to grow more quickly than any other energy source in history.

We are attempting something that has never been done before. In previous energy transitions, new sources did not usually replace existing sources: they simply expanded the mix. The age of petroleum did not mean the end of coal, and neither did the expansion of coal end wood use; indeed, we are today burning more coal than at any other time in human history. Many governments favor this transitional model, known as an "all of the above strategy," where renewable energy grows as energy consumption expands across the board.

But we cannot underestimate the challenge. Energy is not merely electricity. It means supplanting entire infrastructures of transportation and production. Countries such as Uruguay, Iceland and Albania boast huge rates of renewable electricity generation, around 99 percent. But in many states, electricity represents only a small portion of all energy used.

To prevent 2°C temperature rises, while keeping up with current energy demand, requires adding 1,100 megawatts of zero-carbon energy a day to the global energy mix. We are only adding around a tenth of this.[74] Many environmental advocates and politicians have found promise in the possibilities of an economy entirely powered by renewables. Engineering studies have outlined how we can translate current energy needs into a system supplied 100 percent by renewable energy.[75]

While these are laudable and much-needed attempts to defy a fossil-fuelled present, they often assume we can simply insert a very different type of energy into a world geared for fossil fuels. Renewables cannot simply replace fossil fuels in societies geared around the expectations and demand curves wrought by fossil fuels. Our industries and technologies have been designed for the intense combustive power of coal, gas and oil.

Just as gravitational energy differs from heat energy, or mass energy from radiant energy, the energy produced by fossil fuels and renewables is not equivalent or easily comparable. Solar panels are attempts to replicate photosynthesis. Wind turbines are modernized windmills. These are very different types of energy from fossil fuels, with different types of energy quality and levels of energy intensity.[76]

Renewables cannot replicate the same energy density as fossil fuels. When the sun doesn't shine, or the wind doesn't blow, less energy is produced. Renewable sources are fluctuating and intermittent, not always available. This intermittency poses problems for attempts to simply replace fossil energy. Without questioning our energy system as a whole—how much we need, what we use it for, and when we use it—we will simply be trying to square circles.

Many proposals for change effectively attempt to maintain a fossil-fuel world without their defining inputs. The problem of intermittency and baseload energy is addressed through the construction of enormous battery storage facilities, spectacular hydroelectric projects, biomass incinerators and a radical expansion of transmission infrastructure.[77]

Rather than conducting modeling gymnastics to prop up the world of today, we need to start talking about radically shifting our energy culture. That shift will fundamentally involve moderating our endless thirst for energy. We cannot have infinite renewable energy. All renewable technologies require mined metals, embodied energy, and access to territory, whether on land or sea. These material constraints suggest that, as we work to reroute our economy towards renewables, we should also strive for energy sufficiency, measures that will transform how much energy we use.[78]

Our world has both energy extravagance and energy poverty. The average North American uses over twice the energy of the average European, and 30 times the energy used by the average Bangladeshi.[79] The improved welfare of billions in the world relies on increased energy access, but increasing that access within the bounds of what's ecologically desirable and energetically feasible relies on reducing demand on the other end of scale. We should strive for adequate and affordable amounts of energy that meet our needs and that do not harm us, our fellow citizens or the climate at large.[80]

To accomplish this goal, we need a range of measures, from the transport sector to the area of grid management. All major sectors, including heat, have to be electrified, and that electricity has to be sourced from renewable energies. Given the intermittency of renewable electricity generation, load-balancing policies can synchronize energy use with energy production.

Tackling energy waste and improving energy efficiency, the cheapest and arguably easiest measures, also increases the potential to reduce emissions and energy spending. Through the encouragement of shared services—from home deliveries to community kitchens to public laundrettes to public modes of transport—we can dilute the need for appliances and intensified energy use in every individual household.[81]

But we cannot assume that solar power and electric cars will by themselves deliver a more equal world. Let's recall the metaphor of explosives and detonators. A logic of pure renewable solutions might help accelerate the fight to remove the detonator, but can continue loading explosives. What may appear to be a solution may end up entrenching a community's vulnerability. Unless the development of new forms of energy follows principles of justice, we risk recreating the injustices of the fossil economy.

First, this involves recognizing that fossil fuels are historically wound into the economies of thousands of communities. The Kuzbass, Donbas, Silesia, Jharia, the Ruhr, are among the many regions intertwined with coal mining over decades. Any transformation needs to be sensitive to the downstream consequences on workers, communities affected by operations, shareholders and the environment.

Second, it involves recognizing the importance of applying broader principles of sustainability to the expansion of the renewable energy industry. The renewables revolution requires an enormous amount of wind turbines, solar panels, electric wiring and battery storage systems. All these have requirements on energy, materials and land. Large land areas are needed to host heavy-duty renewable projects. Replacing a 1,000-megawatt coal-fired power station requires 500 large wind turbines.[82] Already new battles for the lands of solar and wind energy are emerging, as companies and states set out to meet urban energy demands through larger projects in rural areas.

Desert ecosystems, from the Atacama to the Mojave, are being handed to utility-scale solar projects.[83] The Sahara has been the proposed site of the Desertec project, a large-scale renewable energy-generation plan designed to provide electricity for Europe. Daniel Ayuk Mbi Egbe, from the African Network for Solar Energy, noted: "Many Africans are sceptical about Desertec. Europeans make promises, but at the end of the day, they bring their engineers, they bring their equipment, and they go. It's a new form of resource exploitation, just like in the past." Tunisian trade unionist Mansour Cherni added doubts: "Where will the energy produced here be used.... Where will the water come from that will cool the solar-power plants? And what do the locals get from it all?"[84]

What will be off-limits? We must be wary of the history of environmental trade-offs. As geographer Maria Kaika reminds us, "our sustainability is someone else's disaster."[85] In the 1960s, various environmental groups in the US conceded the construction of the Navajo Generating Station, a major coal-fired power plant, essentially in exchange for cancelled dams in the Grand Canyon.[86]

There are abundant arguments for the rapid expansion of renewable energy. But in our efforts to advance low-carbon energy, we should be transparent about the costs. Implemented without care, renewable energy could exacerbate social inequalities. Solar farm projects have already involved conflicts over forest clearing. In Mexico's Yucatan Peninsula, local communities have opposed corporate wind farm projects in the Isthmus of Tehuantepec over land-rights violations.[87]

Silicon, indium, steel, concrete, aluminium, nickel, copper and tin, cobalt, coltan, lanthanum, lithium, tellurium: these are all resources crucial for making solar panels, turbines and batteries.[88] Wind turbines and electric cars often require rare earths such as neodymium and dysprosium. In the Chinese region of Boutou, where two-thirds of global rare-earth extraction takes place, mines are responsible for intense pollution, farmland destruction, rural displacement and health impacts.[89]

Lithium mining has brought pollution to communities in Tibet, Chile and Argentina.[90] Half of global cobalt is sourced from the Democratic Republic of Congo, in mines dogged by documentations of dangerous working conditions, child labor and malpractice.[91] Global electric car companies have already faced scrutiny from human rights organizations over their supply chains and their own labor practices.[92]

Instead of simply proposing energy transitions that sound promising, we need greater energy literacy to understand how we can actually shift the system effectively and equitably. Well-designed just-energy transitions can mitigate climate change, support workers, cut energy bills, clean up polluted sites, create secure jobs, regenerate rural areas and nourish community life.[93]

Such a transition could also entail a transformative shift in global wealth and power. While fuels and underground deposits can be easily controlled, it is harder to own the sun, wind or waves. Renewable energy could lead to energy democracy—where every community, every region can have its own power stations. Energy, the most profitable business of the 20th century, could become a community right and not a commodity.

Such transitions are already partly under way. The contemporary fossil-fuel industry is facing an existential crisis, as technological, legislative and economic trends pull at the carpet beneath its feet. Major petroleum firms risk having both unprofitable and un-burnable portfolios, as renewable alternatives displace fossil-fuel demand, and nation states introduce bolder climate legislation.[94]

Oil-sector protagonists are admitting a horizon of restraint. In early 2017, the *Wall Street Journal* wrote: "A new era of low crude prices and stricter regulations on climate change is pushing energy companies and resource-rich governments to confront the possibility that some fossil-fuel resources will remain in the ground indefinitely."[95]

But what shape this transition takes, and whether it can genuinely reduce consumption, is up to us.

Dismantling hierarchy

Millennia of injustice have calcified hierarchies between humans, unevenly distributing advantages, disadvantages and burdens along the lines of race, gender and ability, among other attributes. Today, intense inequalities pervade all spheres of life, from the public realm to the domestic. These inequalities are not just momentary phases to be overcome, but rather deeply rooted structures that need to be actively removed.[96]

Many of these hierarchies share common roots. Ecofeminism, for example, is a lens which can help us see connections between environmental and gender justice. Ecofeminist thinking has elucidated how the deeply rooted oppression of women is interwoven and intimately linked with the exploitation of nature.

While conceptions of gender across countries and cultures differ, we have generated widespread cultures of extreme masculinity that valorize belligerent force and deny vulnerabilities, whether of the Earth or its inhabitants. These masculinist visions have shaped our science, validating violence in the conquest of knowledge, and attempting to exert godlike control over nature.[97] Subjected to such logics, our bodies and territories bear parallel brutalities.[98]

Yet all fault lines are fertile for healing. First, we must position the voices and visions of the routinely neglected as central in redressing the past, crafting the present and imagining the future.

Through legal, institutional and cultural changes, we can work to dismantle the inequalities, oppressions and injustices that hold back people's dreams and capabilities. The implications of such maneuvers can ripple across the board.

After all, the lack of parities of power corrode and distort societies. Epidemiologists have found significant relationships between equality and societal well-being.[99] More-equal societies have healthier populations, higher social mobility, less addiction, lower crime rates, lower infant mortality rates, less divorce and less violence.[100]

In ecological terms, we need to pay close attention to those inequalities that lay the explosives for climate violence, ensuring that all responses, recoveries and relief processes are attuned to address vulnerabilities. In an environmentally precarious future, confronting inequality, in all its forms, will save lives.[101]

We also need a repayment of the climate debt, a reversal of the inequalities that continue to drive the climate crisis. Various civil-society organizations have pushed for countries to be given their fair share: a determined amount of emissions reduction that corresponds to the capacity and historical responsibility of states.[102] Those most able to adapt, and most responsible for the threat of climate change, should be those at the forefront of providing resources. Climate finance from the richest to the poorest countries will have to be increased.

Care

Care is the foundation of life. As biologists Humberto Maturana and Gerda Verden-Zoller write: "Love is the grounding of our existence as humans.... It is as loving, languaging beings that we can still become aware of what it is to be a human being, and it is only as loving animals that we can still create the conditions for the upbringing of our children."[103]

We are all dependent beings, reliant on ecosystems and our fellow beings for nourishment, support, contact and companionship. Love, and its bonds, makes life possible and bearable. Our longevity and life quality are firmly tied with the strength of our relationships.[104]

The vision of a world of invulnerability has rendered the deep psychological needs of human beings invisible. It has wrought societies of solitude and isolation, with too many of us in conditions of emotional deprivation.

Although our countries, economies, societies and personal lives rest on care, it is too often seen as something external. This violent distancing has meant that too many humans, primarily women, do huge amounts of work that goes undervalued and unpaid. Affective labor in our societies is offloaded to those with less power, either left to the household or transferred to precarious workers.

We have to stop seeing the care work performed by humans and ecosystems as inexhaustible and invisible. Feminist, anti-racist and ecological economists have instead urged us to rethink economics and the organization of social life by putting life and the satisfaction of needs at the very center.[105]

A society that values care is a very different society, particularly when it comes to the topic of time. Time is the ultimate finite resource. In our industrialized world, where production reigns supreme, days are structured not around the meeting of needs, but around our offering of time to schools and workplaces. To survive, most human beings must sell their time to the labor market. Wracked by time poverty, too many of us lead hectic lives, starved of the hours we would like to devote to meaningful pursuits. Our societies venerate overwork and time sacrifice, with overwhelming workweeks becoming routine. Exhaustion is structured into our societies. This excess of work is contributing to severe health risks.

Under principles of care, the organization of time should align with human needs, not industrial schedules. We can reduce working days, redistribute hours to integrate unemployed people and reshape sociocultural norms to recognize and more equitably share care work.

Instead of cultures of endless work and time extraction, we can start weaving a culture of slowness and stillness, with different rhythms of life. The freeing of time can allow for the freeing of possibility. We can open time to share domestic work, to learn, to rest, to listen, to meditate, to create, to participate in society, to volunteer, to pursue passions and lead healthier lives.[106] Such shifts could allow for greater equalities of gender as well as healthier societies. Reduced work hours have been widely linked with higher productivity, increased mental health, lower blood pressure and better academic performance.[107] Countries with lower working hours have happier citizens and lower environmental impacts.[108]

All this is eminently possible. In 1930, the economist John Maynard Keynes speculated that technology would reduce the workweek to 15 hours by 2028. But while the technological innovation Keynes envisioned allows for this possibility, the lack of sociopolitical courage and innovation has removed it from the public imagination.[109]

Restoration

Our assault on nature has not only accelerated the torrent of emissions, but it has also removed many of the ecosystems that might have protected us from the furies of climate change. The mutilation of much of the world's life calls for a process of major ecological restoration, an extension of care to our lands and polluted waters.[110]

To repair the injuries wrought by careless industrialized development, we must begin to recast our relationships with our surroundings, recognizing the rich depth of ecosystems. Forests are our planet's lungs, its stewards of rivers, its purifiers of water. They restrain erosion, stabilize land, shape weather, replenish water vapor and maintain our climate. Coral reefs are not just picturesque seascapes, but sifters of sand, nurseries of fish, buffers against storms and the basis for atolls. Mosses cleanse water, host countless organisms, build seed beds and compose soil. Sand, voraciously consumed by the construction industry, is crucial for filtering water and maintaining its levels.

The comprehensive ecological roles of all environments show just how reliant we are on their contributions to the web of life.

Ecosystems hand us the gifts of pollination, carbon absorption, soil fertilization, medicinal provision, shade, protection against erosion, water purification and storm protection, among dozens of other gifts. One quarter of prescription drugs are derived from forests: quinine for malaria, curare for anaesthetizing patients in surgeries, periwinkle for a variety of ailments.[111]

Given the depth of their contributions, every case of ecological destruction is a crime of many dimensions. Deforestation, for example, is not just the removal of trees. It is the destruction of ecosystems that manage and stabilize water cycles. It is an invitation to erosion, as rains carry soil downstream, increasing the silt density of rivers, which raises the risks of floods. It is the direct exposure of lands and their inhabitants to the sun which elevates temperatures, heats surfaces and desiccates the soil below.

The constant removal of ecological defences such as corals, mangroves, marshes, forests and wetlands makes land more vulnerable to extreme weather. Ecosystems are the first lines of defence against storms, acting as breakwaters. In ecology, a simple rule generally prevails: the healthier the ecosystems, the lower the climate risks.

By actively rewilding, regenerating and restoring ecosystems, we can increase our resilience, reverse the loss in biodiversity and protect livelihoods. Initiatives to sustain carbon sinks, from foresting to regenerating abandoned farmlands to protecting peatlands, could also contribute significantly to emission reductions, absorbing billions of tonnes of carbon.[112]

Yet restoration or planting are poor substitutes for protection. Tropical trees, for example, amass up to half of their biomass in the last third of their life. Old-growth forests capture far more carbon than newly planted or degraded forests.[113]

We protect what we depend on. Those dependent on forest environments for water, food, medicines and shelter have the greatest stake in them. One of the most effective ways for protecting forests is delivering them into the hands of those communities who rely on their safety, and have thrived in proximity to them. From Brazil to Bolivia, Nepal to Nicaragua, countless studies have found that community forest concessions have been remarkably effective in controlling deforestation.[114]

Although indigenous territories and protected areas represent over half of the Amazon's forested land, only 17 percent of the region's deforestation has occurred in this area.[115] In the Peruvian Amazon, deforestation plunged by 75 percent in territories once they were formally passed to indigenous communities.[116]

Embracing community management means challenging dominant thinking about conservation, which has conceived of protected areas as territories emptied of human life. Safeguarding ecosystems has come to mean separating them from humans. Thousands of forest-protecting communities have been evicted to make way for visions of environmental protection. In Thailand, Hmong and Karen hill tribes have been designated as "illegal occupants" of their ancestral land. Baka communities in southeastern Cameroon have faced beatings and murders by conservation officers. In Botswana, Kalahari San people have fallen victim to a shoot-to-kill policy instituted to protect the Central Kalahari Game Reserve. Studies across dozens of protected areas in Africa have documented endemic human rights abuses and displacement of thousands.[117]

Adaptation

Our current social systems are incapable of handling the future world as projected by climatologists. If there is warming we cannot avoid, how can we reduce its impacts?

Adaptation involves prioritizing the alleviation and prevention of suffering. It asks: how prepared are our societies to greet what will come? What infrastructure will meet the weather of tomorrow? Are responses to extreme weather left to governments or to the market? What are our defence lines for disaster? Who gets access to housing? Who is prioritized in recovery plans? How do communities shape decisions about their fates?

Adaptation is the reduction of vulnerability in the face of disaster. It can mean physical infrastructure: sturdier homes, expanded defensive barriers, fortified electrical grids, strengthened water pipes, installed early-warning systems, improved meteorological equipment, hospitals with back-up energy systems. It can mean ecological infrastructure: planting trees to shore up erosion, protecting mangroves that buffer against storms, improving a city's green cover, replacing impermeable paved surfaces, using new crop varieties more resistant to extreme weather. It can mean educating people from an early age about environmental risks, improving disaster-response plans, coordinating evacuation routes, opening treatment centers, establishing emergency shelters and designing recovery plans that work to make sure people have lives to return to.

Robust adaptation measures have brought death tolls and injury rates down across the world. In India, a major typhoon killed over 10,000 people when it struck the state of Odisha in 1999. Over the succeeding years, major changes were put in place. Response plans were drawn up, effective warning systems installed and cyclone shelters arranged. When an analogous storm, Cyclone Phailin, rammed into the Odisha coast in 2013, around a million people were evacuated, with the death toll brought down to 18.[118]

In Cuba, measures taken in the wake of 1963's devastating Hurricane Flora were instrumental in preventing fatalities over the next decades. A new plan, named Project Life, prohibits construction in vulnerable coastal areas, mandates relocation for communities and moves crop production away from areas vulnerable to saltwater contamination.[119]

In the Indian city of Ahmedabad, a heat action plan reduced heat wave deaths from hundreds to dozens between 2010 and 2015, by improving public awareness, coordinating between government agencies, training healthcare professionals and formalizing prevention methods that mean when heat waves arrive, cooling centers are opened and water is automatically delivered to slum communities.[120]

Successful adaptation initiatives have had broader implications. In the Ethiopian region of Tigray, fruit nurseries, health centers and creative irrigation schemes have helped improve social strength before droughts. In the Netherlands, bold urban design projects have incorporated public amenities as receptacles for floodwater. Basketball courts, parking lots and public parks double up as emergency reservoirs for sewage overflow.[121]

Rather than seeing adaptation as coping or adjustment to disaster, we need to see adaptation as a transformative process that can tackle injustice, reduce social vulnerability and build resilience.[122] Assembling physical defences is as important as the social relationships in any neighborhood. Detailed studies of the impacts of heat waves across Chicago in 1995 found contrasting conclusions: 8 of the 10 most affected areas were the poorest parts, largely populated by African-American residents. Yet simultaneously, three of the neighborhoods with the lowest death rates were also some of the poorest parts, largely populated by African-American residents. Neighborhoods with strong community institutions fared best.[123]

Just political infrastructures are also indispensable. What kind of decisions are taken in the wake of a disaster? Whose relief is prioritized? Democratic and community-based approaches to adaptation can allow for responses that build collective strength.[124]

Migration is also intimately tied up with adaptation. In a future of climate violence and its associated displacements, we must work for a world that values the protection of human beings before the protection of borders.[125] Justice is a double-game. We must defend both the right to move and the right to stay. The right not be displaced by a violent climate's coercions and the right to find sanctuary in the wake of displacement.[126] Currently, those displaced through environmental changes live in a legal limbo, with precarious rights or recognition. On an international level, there is a need for strong international principles, protocols and protections that protect communities forced out of their homelands by climate change and allow migrants safe options for movement, entry and citizenship.[127]

Education and culture

Education is the transmission of memory. It hands down the legacies of the past, showing us what is important and valued. But decades of restrictive education have created a learning of separations. Our knowledge is partitioned and fragmented, with students led down the slender tracks of humanities, sciences or arts.

We prioritize abstract, text-centered learning that denies our ecological connections and distances thinking from doing. The education system too often neglects our imaginations, our spirit and our emotional intelligence. Collective creativity and cooperation are stunted, as students are subjected to constant comparison and competition. Success is associated with outperforming the other.

Educational systems bear the mechanical influence of industrial planning, divided into strict units and lesson plans, with examination and quality control at the end.[128] Students are scored, standardized and tested, their confidence damaged by established hierarchies. Grades, diplomas, degrees and certifications enshrine distinctions that will bear their mark for decades. Education becomes a process of accreditation, and preparation for professionalism, as young people are encouraged to define their career paths and monetize their aptitudes.[129]

Learning is regimented and dictated, leaving students with little time to play to discover the world around them and who they are. Children become projects, trained and formatted for the job market. Our curricula and mainstream textbooks are suffused with technological optimism, with scant attention paid to the deep ecological challenges young generations will face.[130]

Although more children are attending school around the world than before, that is not necessarily translating into high-quality education. School construction and subsequent attendance doesn't always lead to learning. In the Indian state of Andhra Pradesh, only five percent of children in fifth grade can do basic arithmetic.[131]

In too many contexts, young people can recognize dozens of logos, yet few species. Cognitively distanced from the webs of life that sustain us, we have little idea about how food is produced, where water flows from, or where our waste ends up. We are profoundly illiterate in ecological terms. Literacy is ultimately our ability to read the world. The Brazilian educator Paulo Freire saw illiteracy as a way of closing our doors to reality and our surroundings. Care is rooted in attention. We need educational and cultural solutions that heal the rift between nature and culture, and improve our ecological literacy: our capacity to interpret ecological systems; our ability to know how life works. [132]

We need an education of care, which teaches us to tend to vulnerable bodies, minds and ecosystems. An education that nourishes our creative confidence and teaches us to healthily manage and use the commons. An education that sheds the boundaries between disciplines, and helps students see connections and relationships before specializations. Rather than solely acquiring analytical skills, we need to learn more from life: farming, foresting, cooking, upcycling, designing, healing, teaching, communicating, performing, storytelling and building.[133]

We also need an education of memory, teaching a critical and reparative history of the world—an education which fosters ecological citizenship, taking learning out of the classroom and into our ecosystems and communities. In India, the political organization Swaraj Abhiyan launched an initiative of drought duty, allowing urban students to spend periods of time in drought-hit villages.

Ultimately, we need imaginative, empathetic, critical and creative thinkers for the 21st century, responsible for the world. Neither should education be restricted to our younger generations. We can amplify the terrain of possibility: solutions on buses, on streets and billboards. Climate-change warnings plastered on petrol stations. Educational weather forecasts. We can establish community learning centers, serving as hubs for learning and knowledge creation. We can establish food centers that teach nutrition, recover culinary memory and encourage people to be more conscious of what and how they eat. We can transform medical institutions into centers of holistic health.

Facts alone won't switch behavior. Our identities are shaped by our broader cultures. Our monuments, curricula and anthems feed us national histories and founding myths.[134] Traditions, ceremonies and rituals build habits, shape norms and determine what is valued. Our desires, wants and perceived needs are very much shaped by a commercialized culture that associates well-being with status.

Many of our dominant ceremonies, from weddings to birthdays, focus on us as individuals. Rarely do we acknowledge our collective dependence, and our ecological reliance on the land around us.[135] Rather than festivals of consumerism, such as Black Friday (United States) or Singles' Day (China), we could attach more value to commemorations that inculcate wonder, and encourage ecological contact and collective joy. These already exist in plenty, marking solstices, harvests or the coming of winters. In Iran, Sizdah Be-dar, the Persian Festival of Nature, is an important moment in the annual calendar, where people are encouraged to join nature and spend the day outdoors. In Judaism, Tu B'Shvat marks the New Year of the Trees, a day to acknowledge the debt of humans to nature.

In addition to culture, the media act as a powerful filter in influencing our perception of reality. The distance and unimportance we ascribe to ecology is partly derived from the exclusion of environmental issues from public conversations. We lack an apparatus of attention that draws connections and makes them visible. According to the Red Cross, over 90 percent of what are known as "global crises" go unnoticed.[136] Protracted crises are neglected. News cycles, in the desire for breaking spectacle, jump from outrage to outrage. Every disaster is treated like a cataclysmic novelty, its systemic roots blotted out. We need art and broadcast media that are attentive to a world of climate violence, ready to humanize its expressions and illustrate its resonant links.

We need to confront the silence about ecology, and the enduring deafness to the voices of those on the frontline of its impacts. The visions, solutions and alternatives crafted in the Majority World are dramatically underrepresented. Although a significant part of the global population draws its livelihood from the land, our media is largely silent about agricultural issues. Rural and farmer voices are largely absent from print, visual and social media, leaving the agenda to be set by urban elites.

Finally, creating a society that confronts the challenge of climate change means shifting the influences that promote ideals of over-consumption. Our consumer societies, awash in advertising, fan feelings of envy and status anxiety. The "good life" is tied to conspicuous consumption. Weaving together a more ecological society and economy will be implausible without tackling the invisible curricula of advertising, increasingly dominant in a digital age. Former Google designer Tristram Harris described "advertising" as the "new coal," as it "got us to a certain level of economic prosperity, and...it also polluted the inner environment and the cultural environment and the political environment because it enabled anyone to basically pay to get access to your mind."[137]

Cities such as São Paulo and Grenoble have outlawed outdoor commercial advertising through bold campaigns to confront visual pollution.[138] Without changing who we listen to, what we are exposed to and whose voices we value, we will struggle to reach climate safety.

Healing

Nature is the basis of human life. Our health rests on clean air, drinkable water, nutritious food and safe shelter. In English, the word "health" finds its origin in the Anglo-Saxon word for whole: health is only attainable when whole.[140]

We must work to foster healthier societies and people, across physical, mental, nutritional, communal and environmental dimensions. Human health relies on the health of our surrounding ecosystems, and the health of ecosystems relies partly on the health of one of its core members, human beings.

Our modern traditions of medicine, and their subsequent institutions, have focused predominantly on individual health. Personal prevention takes precedence over systemic work. Immense amounts of research have worked to pioneer surgical technology, pharmacology and specialist care. While these advances are vital, this model is frequently inattentive to the systemic causes of poor health.

Poverty, exploitation, the pressures of a competitive culture, exclusion, loneliness, injustice: all of these disappear under the label of depression. Asthma prevention focuses on individuals reducing their exposure to asthmagens, rather than on the wider causes (air pollution). Treatment of cancers relating to pesticide exposure rarely includes communal measures to reduce or eliminate its use. Diagnoses of obesity, poor diet and stress largely ignore their social drivers.[141] Health departments, medical practitioners and epidemiologists primarily focus on the proximate causes of diseases, narrowing their prescriptions to what individuals can do to ameliorate their health.[142]

Such a focus coincides with our broader culture, which paints poverty, hardship and injustice as individual failings. All problems are endogenous, all weaknesses personal. Societal problems are internalized as personal deficiencies. Our sickness, our sadness and our despair are of our own making.

Yet all behavior is determined by our environments and by wider interactions. Various physicians and medical practitioners are calling for more ecological approaches to health that are attentive to such connections.[143] Rather than simply seeking the cure for a patient's personal illness, healthcare practitioners are encouraged to look at the health of a person's context: the health of their physical, social, environmental and political environment. Public health becomes not just the treatment of illness, but its prevention: the removal of conditions for bad health.

Ecological and public health solutions coincide. Restoring ecosystems and preventing pollution can both extend lives by many years and support climate safety. Exposure to rich ecosystems has been found to bring substantial health benefits, from improving psychological well-being to supporting physical therapy. Greener urban environments have been associated with reductions in violence, and more egalitarian child play.[144]

Reconstruction: Cities and space

Concentrating people, consumption and possibilities, cities are epicenters in the struggle for climate justice. Although cities make up just two percent of the planet's land surface, they are vital to the global metabolism: three-quarters of all goods are consumed inside them. Urban areas bear enormous footprints, ingesting copious amounts of materials and disgorging masses of waste. Ten years ago, the proportion of people living in cities overtook those living in rural areas. By 2050, urban dwellers are expected to represent over two-thirds of the world's population.

Oriented by the logics of modernity, urban planners over the past 100 years have imposed blueprints on the complex ecology of cities. They have arranged cities of strict separations, divided into functions: residential quarters, industrial parks and cultural centers. The distances wrought by separation were quelled by the logic of motorization, as public space was designed to prioritize the circulation of cars. Knotworks of roads and highways were stretched over territories. Today this motorized transportation system is the leading source of emissions in many countries. Astounding levels of road violence cost 1.3 million lives each year. To add insult to injury, even more people die from exhaust-pipe emissions than from car accidents.[145]

Cities, conceived as landscapes free of nature, sprawled by consuming territories. Wetlands, forests, farmlands, marshes, lakes and rivers have been cleared for the advance of expansive metropolises and megacities. Buildings have been developed along waterways or floodplains, with paved roads covering cities in impermeable surfaces.[146]

Today urban centers are reaping the sorrows of this model of modernity. In our cities, we have overlapping crises of congestion, pollution, poor health, overcrowding, gentrification, unaffordable housing and loneliness. Choked by cars and suffocated by fumes, the air of many cities has become unbreathable, and chronically hazardous for their residents. Too many citizens live in dangerous and depressing living conditions. Unequal urban development has crafted enclaves of wealth and poverty. Up to 80 percent of African city dwellers live without secure property rights, haunted by insecurity and the threat of evictions.[147]

New visions of a more intricate and sustainable urbanism are emerging, which view cities as ecosystems and spaces for life. As complexity scientist Geoffrey West explains, the purpose of these visions is "finding a way to minimize our distress while maximizing our interactions." Territories are destined for mixed uses, combining residence, work, services, agriculture and recreation. New relationships between nature, home, work and play are allowed to emerge. Public space devoted to automobiles is opened to foster neighborhood life and support pedestrian or cycling activity.[148]

Through such transitions, our cities can be developed to be healthier, safer, more prosperous, more inclusive, more accessible, more diverse, more livable, more equitable and more reflective of our values. Urban planning can embrace ecological thinking, developing more breathable spaces, ensuring cities are more resilient to extreme weather, and opening up room for biodiversity and other beings. Rather than uniform cities, we can recover vernacular architecture and traditional methods of construction adapted to local climatic conditions.[149]

If we mix up the city, strengthen its neighborhoods, and reclaim space for citizens, we can encode justice into the city's fabric, making daily living healthier and lighter.[150]

CHAPTER 14

WHAT THEN MUST WE DO?

*Ecology without social justice is
little more than gardening.*
ENRIQUE VIALE

▪ ▪ ▪ ▪ ▪ ▪ ▪ ▪ ▪ ▪

We are not drowning; we are fighting.
Slogan of the Pacific Warriors

▪ ▪ ▪ ▪ ▪ ▪ ▪ ▪ ▪ ▪

*The standard of justice depends
on the equality of power to compel.*
THUCYDIDES

▪ ▪ ▪ ▪ ▪ ▪ ▪ ▪ ▪ ▪

A S THESE DIVERSE VISIONS show, our problem is not the absence of alternatives, but the absence of meaningful action. Yet the magnitude of the efforts necessary can leave us stranded in impotence.

Many of us stumble at the very first question: what can I do? But our failure to find a straightforward answer lies partly in our misdirected question: there isn't much that individuals working alone can do. But together, as participants in society, there is a lot we can do.

For too long, the onus of tackling climate change was loaded on the backs of individuals. Campaigns, textbooks and documentary films called on us to act in our own lives. We were told that by changing our consumer habits, by recycling, by switching our lightbulbs, by shortening our showers, by eating less meat, we could shortcut our way to climate safety.

This approach is certainly important in encouraging personal responsibility. Individual footprints are not inconsequential. Every burst of fossil electricity, every liter of fuel consumed, every flight taken, becomes an equivalent mass of gas in the atmosphere.

But this focus on individual actions can blind us to the structural causes of problems and to the factors that restrict personal behavior. Careless narratives of blame and shame can also be immobilizing, discouraging people who are already exhausted and demoralized by our broader culture.

As journalist Martin Lukacs writes: "Neoliberalism...tells you that you should not merely feel guilt and shame if you can't secure a good job, are deep in debt, and are too stressed or overworked for time with friends. You are now also responsible for bearing the burden of potential ecological collapse."[1]

Ultimately, this is a story which confuses people, telling us that a gargantuan civilizational crisis is soluble through simple personal choices. The only hope of securing the bold transformation required by climate violence lies in collective responses, for we are intimately dependent on each other's actions for survival.

Seas, seeds, rivers, emissions, soils and forests do not obey the carvings of borders. What does the national interest represent when storms barrel across continents? Wind currents will deposit the air pollution of one province into the lungs of another province's citizens. The actions of the world's largest state have the potential singlehandedly to deplete the global carbon budget and radically alter the trajectory of climate violence. The absent boundaries of our environments force us into solidarity.

In the 1970s, the US National Academy of Sciences reflected in a report: "The resources of all countries should be regarded as part of an interdependent habitat rather than merely as possible sources of supply...our national policy should therefore conform to the principles of conduct adopted by the community of nations in a common effort to protect the human habitat and its resources."[2]

Climate violence surpasses our own experiences and personal loyalties. It urges us to converge on all levels: local, municipal, national, regional, international. Balance is essential. If we focus too much on the global, we miss the particular. If we latch ourselves to the local, we lose the breadth of our problem. The poet Rumi spoke of living like a drawing compass, one leg rooted in the religion of birth, the other spinning across nations.[3]

Stepping into the sea

How do we act within complex systems? What decisions do we take, what moves do we make, to push the world into a better direction?

The world is like a turbulent and unpredictable sea, riven by countless connections. Any action or intervention will have anticipated and unanticipated ripples.

This map of political reality is at odds with many conventional pictures. Political actors talk of engineering change, viewing the world as a pliable machine or a geopolitical chessboard. Social change deploys the language of military strategy: campaigns, frontlines and tactics.

And when expectations of change fail to materialize, we ask: why did this auspicious government fail? Why did this progressive initiative collapse? Why does the promise of change rarely become its fulfilment?

To understand political failure means ridding ourselves of the illusion of political control.[4] Politics is not choreography but the opening of possibilities. As activist Saul Alinsky wrote, "there are no rules for revolution any more than there are rules for love or rules for happiness."[5]

So how do we unlock the change we want to see? Do we encourage individual behavioral change or strong social movements? Do we pursue legislative endgames and bold electoral campaigns, or do we support frontline resistance or mass revolution? Do we work within the existing system, or do we exit the system to create new spaces? All these questions, and their associated answers, reflect competing theories of change, different ways in which people understand how transformation can come about.

There are many legitimate approaches, but understanding the obstacles we face is central. Huge profits have been made from the establishment and endurance of the status quo. A lot of people have a lot to lose from the solutions to climate change. These are often called vested interests: individuals or groups who have strong reasons to protect their money and power against attempts to tackle the ecological crisis. From Saudi Arabia to Sudan, countries draw significant wealth and power from controlling energy. Any glance at a list of the world's most valuable companies will see a swathe of fossil-fuel and utility companies. For such interests, action to address climate change represents a massive surrender of power and wealth.

Our innocent intuitions may lead us to hope that these powers may come to terms with the challenge and act accordingly. From a young age, fairy tales feed us faith that the bullies will lose. But politics is about strength, about the hydraulics of new powers displacing old ones. Can we unfurl a transition of sufficient velocity to really change the world? Can we assemble enough power, in enough time, to stop catastrophic climate violence, to enable communities to face the coming furies, to ensure that we do not become a divided world of fortresses?

To set in motion the tectonic shift in power needed to tackle climate violence, we will need to array forces like never before. The case for transformational change will have to gain mainstream relevance and public support. Our pervasive high-carbon economic infrastructure, predicated on the use of fossil fuels, must be remodelled at a pace unseen in human history. Our nature-gorging economies, and the worship of indiscriminate growth that drives them, will have to be boldly questioned. The abysmal injustices and inequalities that foster vulnerability will have to be addressed through comprehensive reparations, healing, education and economic programs.

To shake up the inertia of the status quo, we will have to organize, mobilize broad bases, popularize visions and make what is currently marginal into the new common sense. We must be tactically innovative. Too often, social change is conceived through the prism of our old system and its existing institutions. We need to be both defensive and future-oriented, resisting the undesirable while rebuilding the new.[6]

The repertoire of our tactics should be broad. We should build electoral alternatives that will ignite people's imaginations. Creative mass protest and nonviolent disobedience can escalate activities, shift parameters, create defining moments and unlock changes.[7]

We need to channel efforts into strategies that transform possibilities, realize alternatives and build widespread support for economic transformation.[8] It is an enormous task, but it is far from impossible.

First, as political scientists Erica Chenoweth and Maria Stephan document, the historical evidence shows that successful social transformation relies on the sustained participation of a strong minority—around 3.5 percent of a population—to obtain success.[9]

Second, when we think strategically, we realize there is ample low-hanging fruit around us. Many of the richest states can decrease their material intensity and electricity use significantly without drastic changes.[10] Even simple changes with lower levels of resistance, such as collapsing the emissions of refrigeration systems, could have significant impacts.

If 10 percent of the global population, which is responsible for half of all emissions, reduce their footprint to that of the average EU citizen, we would see a 30 percent decrease in global carbon-dioxide emissions.[11] Huge emissions reductions could potentially be triggered simply through removing handouts to polluting industries.[12]

Introducing progressive taxes on single-use plastic, deposit reward and return schemes, ensuring minimum standards for the durability of products, issuing compulsory requirements for recycled materials: all of these offer places to start.

Connection and solidarity

Centuries of thinking have bequeathed us with a worldview of separations. But, just as our crises cannot be understood in isolation, they cannot be solved in isolation. Common threads bind together environmental destruction, inequality, racial exclusion, poor health, violence, patriarchy and poverty.

On Christmas Eve in 1967, crowds filled the Ebenezer Baptist Church in Atlanta, Georgia. Reverend Martin Luther King Jr. addressed the congregation:

> Now let me suggest first that if we are to have peace on earth, our loyalties must become ecumenical rather than sectional. Our loyalties must transcend our race, our tribe, our class, and our nation; and this means we must develop a world perspective.... It really boils down to this: that all life is interrelated. We are all caught in an inescapable network of mutuality, tied into a single garment of destiny. Whatever affects one directly, affects all indirectly. We are made to live together because of the interrelated structure of reality.[13]

In contrast to these connections, our societies are cordoned into tidy silos. People are split into rigid academic disciplines, occupations and roles, and are separated by jargon. Our political movements hand us splintered menus of single-issue campaigns.

We cannot afford to maintain this piecemeal and divisive framework any longer; we need to initiate conversations, from botanists to builders, from architects to artisans, regarding how we can collectively build the common ground of transformation. We need to turn incompatible visions into overlapping aspirations. We need to come up with strategies that focus on suturing the separations, bigotries and privileges that preclude cooperation. As anthropologist Xochil Leyva reminds us, collaboration fundamentally means "co-labor," working together.

With compassion, courage and humility, we can attempt to find understanding. We can thread solidarities where they have been crushed through division, insularity and hostility. We can join efforts, learn across disciplines, braid unlikely affinities, pool resources and bridge our divides.[14]

In Canada, the Leap Project convened First Nations leaders, trades unions, environmentalists, migrant-rights activists and campaigners from a range of backgrounds to start building a common agenda. The results of that process were released as the Leap Manifesto, a bold vision of a world defined by care for the Earth and each other, a world where no one is disposable. Viewing the climate crisis as an opportunity to build a new economy, it puts forward solutions that will save money, improve neighborhoods, address historical injustice, support public health and weave a more equal economy. The Leap is just one of many projects working to assemble broader coalitions.

The challenge exposes the limits of reason. Solidarity cannot be arranged. The poet Yehuda Amichai described these deficiencies in a poem titled "The Place Where We Are Right":

> *From the place where we are right*
> *flowers will never grow*
> *in the spring.*
>
> *The place where we are right*
> *is hard and trampled*
> *like a yard.*
>
> *But doubts and loves*
> *dig up the world*
> *like a mole, a plough.*
>
> *And a whisper will be heard in the place*
> *where the ruined*
> *house once stood.*[15]

Communication

The word "apathy" comes from the Greek *apatheia*: to be without feeling.[16] For many people, climate change as it is traditionally framed is distant and emotionally irrelevant, leaving us anesthetized to the conditions of our changing climate. In other cases, it produces confusion when people want clarity above all. Some research has shown that, for many, attention fades when simple solutions dissolve into complexity.[17]

Climate change's abstract framing, its faraway victims and its intricacy present cognitive obstacles.[18] In many societies, environmental issues are also perceived as elitist, luxury issues. Ecology has been understood and ridiculed as the defence of birds and flowers.

As the linguist George Lakoff has documented, verbal and visual frames imbue the way we think and imagine. Metaphors are pairs of glasses. The environmental debate is defined by the language of measurement and accountancy: costs, tonnes, digits of destruction.

So how should we instead talk about our ecological crisis? The evidence is mixed.

Narratives soaked only in fear and alarm are rarely useful. Despondency lends itself to disconnection, more likely to alienate us than to motivate us. But optimistic messages can make it seem like the risks are less great or our predicament not so urgent.[19]

To reach all sorts of people, we need all sorts of tools and metaphors. Following Arjun Appadurai's recommendations, we need to speak more in the language of possibility (aspirations, dreams), and less in the language of probabilities (costs, benefits, risks).[20]

We can use other vocabularies, especially those monopolized by the dominant ideology. We can frame environmental safety as liberty, and all threats to our land, water, air and bodies as threats to our freedom.[21]

We should also work to connect issues with the daily mundane struggles of well-being that all people face. The Azeri journalist Khadjia Ismayilova once observed in her analysis of state repression: "The government will fail in these attempts to silence everyone, as there is a truth-telling machine in every house—the empty refrigerator."[22]

There are truth-telling machines, in all our daily frustrations. Our conversations on ecological alternatives need to speak to these, addressing the needs of everyday life: piped water, impermeable roofs, the heat of homes, rent payments, waiting times in hospitals, bus fares, the quality of care facilities, the length of commuting times, the safety of streets, the tedium of jobs, time poverty and the price of culture. We will be aware of our success when ecology is deemed synonymous with housing, water, healthcare, fulfilment and well-being.

Ultimately, our success in expressing the scope of climate violence will lie in our ability to understand each other's aspirations, sufferings and possibilities. Can we propose a viable alternative to dissatisfaction?

Ordinary hope

We are not building a new world from scratch. While the news parades the victories of despair, behind the coverage, countless existing alternatives are being forged today. This is a world of co-operatives, farmers' markets, sharing clubs, communal networks, community energy schemes, solidarity kitchens, transition towns, repair cafés, cooperative housing, alternative currencies and community land trusts. Cities, towns and rural villages are filled with ethical banks, credit unions, ecological farms, popular libraries, kitchens turning food waste into tasty meals and orchards nourished through diverted organic scraps.

From Zagreb to Beirut, Valparaíso to Jackson, Bulawayo to Solapur, the luminous seeds of possibility are everywhere. Landless workers are reviving degraded farmland through agroforestry. Farmers are replenishing seed banks, living libraries of plant genes, to safeguard disappearing crops and weave new traits into crops. Dozens of cities are combining adaptation to climate change with more equitable urban policies.[23] Anonymous leaders and unpatented innovators are leading silent revolutions, hidden in plain sight.

Some of the great human contemporary collective achievements stem from peer production, where people collectively create commons through open design. Wikipedia has become perhaps the greatest human archive of information. Zooniverse is a platform where thousands of citizens collaborate with scientists to produce high-quality research. Wikihouse brings architects, engineers, designers and builders together to shape innovative, sustainable construction technologies.

The world over, social movements have overcome staggering odds. From confronting homophobia to denouncing the entrenched culture of sexual violence, silenced issues have been transformed into dominant conversations within years. Thousands of extractive projects have been defeated by communities. Dozens of countries and communities have outlawed fossil-fuel extraction, or declared themselves off-limits from mining.[24]

The global divestment movement has pushed thousands of institutions to withdraw their finances from fossil fuels. Bold energy transitions, even in the heartlands of the coal and oil industries, are under way. DONG Energy used to be Denmark's largest oil and gas company; today, it has successfully sold all its fossil-fuel assets, transitioning into the world's largest renewable energy company. Social movements in countries highly dependent on fossil-fuel revenues are demanding and gradually achieving bold transitions away from oil and coal.[25]

Thousands of cities and municipalities are pushing bold commitments on emissions reductions. Over 1,000 laws on climate change are now in place around the world. Court cases on climate change are exploding, with citizens suing states, utilities and oil corporations.[26] Since 2012, the number of cities in China publishing daily air-quality data has tripled. Thousands of factories have been pushed to disclose their pollution.[27] From New Zealand/Aotearoa to India, struggles to give legal standing to rivers and forests are increasing recognition of the rights of nature. The International Criminal Court has recently expanded its remit, suggesting possibilities for prosecutions relating to environmental crimes.[28]

And on the frontlines of our ecological crisis, life is always striving to live. Repression breeds resistance, and there are few things mightier than the power of pain. Mothers and fathers fight against the pollution of their children's bodies. Communities resist industrial dumping and poisoning. The residents of Andalgalá, Ain Salah, Cajamarca, Imider, Owino Uhuru, among so many others, are all symbols of resistance to models of development that corrode nature and the lives of its inhabitants.

CHAPTER 15

HOPE, A HORIZON

If the Earth [mati] breaks so much,
how do we stay manush [human]?
Overheard by NAVEEDA KHAN, along the Jamuna River,
in rural Bangladesh[1]

■ ■ ■ ■ ■ ■ ■ ■ ■ ■

I feel deeply that I belong to the Earth.
I belong to nature and I believe in solid ground.
I think we are human and don't have any shelter but humanity.
BEHROUZ BOOCHANI

■ ■ ■ ■ ■ ■ ■ ■ ■ ■

Cynicism is the kryptonite of change.
IBRAM X KENDI

■ ■ ■ ■ ■ ■ ■ ■ ■ ■

Maatu hamru, paani hamru, hamra hi chhan yi baun bhi...
Pitron na lagai baun, hamunahi ta bachon bhi.
[Soil ours, water ours, ours are these forests.
Our forefathers raised them, it's we who must protect them.]
Song of Gharwali women, Chipko movement

■ ■ ■ ■ ■ ■ ■ ■ ■ ■

Life united will never be defeated.
GUSTAVO WILCHES-CHAUX

■ ■ ■ ■ ■ ■ ■ ■ ■ ■

I

T IS TEMPTING TO BRING this capsule of words to a close with certainty, leaving a set of strict diagnoses and thundering conclusions. I do not have any of these. If this book has one message, it is perhaps that approaches led by confident simplicity may carry within them the seeds of many of the crises we face today.

Instead, I'd like to end with a few words on hope. Hope is both the recognition of possibility, but also the courage not to confuse today with tomorrow. In this sense, hope is the line we draw in the sand, the standard we set to prevent ourselves becoming inured to injustice.

We have become accustomed to too much: to a violently unequal world, to the systematic degradation of ecosystems and the constant theft of our future.[2] Without the challenge of change, the indefensible blends into the unquestionable. We assume our temporary realities as certainties, and surrender our aspirations to a different future.

The cinematographer Fernando Birri was once asked at a public event, what is the purpose of utopia? Birri thought for a minute before replying: "[Utopia] she's in the horizon. I step two steps closer, she takes two steps back. I walk ten steps, and the horizon moves ten steps further. As far as I walk I'll never reach it. So, what is utopia for? For that—to walk."[3]

As we try to walk, we may feel shackled by cynicism. Our predominant culture of hopelessness leaves people frightened, and despondent, if not antagonistic to change. We are convinced of our impotence, if not our own worthlessness.

The psychologist Ignacio Martín-Baró, working with schoolchildren in El Salvador, noted:

> Day by day they learn that their efforts in school get them nowhere; the street does not reward them well for their premature efforts at selling newspapers, taking care of cars, or shining shoes; and therefore it is better not to dream or set goals they will never be able to reach. They learn to be resigned and submissive...through the everyday demonstra-

tion of how impossible and useless it is to strive to change their situation, when that environment itself forms part of an overall oppressive social system.[4]

This fatalism creates the futility of action. Fossil-fuel companies play on futility, describing our future as irredeemably dependent on fossil fuels, to ensure it stays that way. They portray ambition as unlikely, highlighting the obstacles to any alternative, to fortify them further.[5] These visions of dire impossibility are articulated by many, from environmentalists to military authorities.[6] The notion of cataclysmic collapse is alluring. We can easily develop a penchant for seeing the end of systems in everything, or for constantly despairing about prospects.[7] That kind of perspective, though, not only obscures incremental gains, but also quickly draws us into impotence. Climate models and studies outlining our future depend on such assumptions of indifference. It is up to us to prove them wrong, and to rewrite the human limitations suggested for us.

▪ ▪ ▪ ▪ ▪

Power is the ability to shape the future, to craft our memory. In the right hands, and used in the right way, power can open doors to freedom. It can give a person more ability to define their own life. But, used unjustly, power is perilous.

Our power to shape the world has never been greater, our weaponry has never been deadlier. We have played with enormous amounts of fire. The nuclear revolution broke open the nucleus of the atom, signalling the ability of humans to release one of the strongest forces in the universe.

We have become more powerful, but more impotent to the abuses of our power. It is this immense power, abused by the few, that has unleashed climate change, a bitter testimony to the might of mendacity. The dominant extractive economic system has been so destructive that it has been capable of derailing the Earth system.[8]

We are not helpless. We still wield enormous influence over the future. What human beings decide to do in the next few years will shape human destiny. The actions of the coming decades will shape the next few millennia. Although our power in an enormous universe may feel small, we should remember that complex systems, in physics and biology, have the property of emergence. The smallest entities can generate unexpected changes in larger entities. Simple objects can give rise to complex patterns. Microscopic processes can have macroscopic impacts. New configurations can emerge as old constraints break. All systems always have leverage points that can rearrange possibilities.[9] In nonlinear systems, surprises await us.

▪ ▪ ▪ ▪ ▪

In history, we are neither plague nor pathology, protagonist nor pawn. The world has many sides. We are capable of unwarranted cruelties, of insidious callousness, of constantly eluding truths that hurt. But we also carry immense, spirit-shaking strength.

Humanity is hard to extinguish. Even in the darkest times of our past, the seeds of human dignity survived brutality. Soviet journalist Vasily Grossman, who followed the Red Army and stared into the bleak horrors of World War Two, noted that "Human history is not the battle of good struggling to overcome evil. It is a battle fought by a great evil, struggling to crush a small kernel of human kindness. But if what is human in human beings has not been destroyed even now, then evil will never conquer."[10]

Hope is not about obsessive optimism. It is a commitment to dream and affirm that change is possible. In part, we have a responsibility to hope. But we need a healthier hope that can be sustained in the face of relentless tragedy. We will not prevent everything. Tremendous pain is inevitable, seared into the future. But our actions can keep worst-case scenarios in the realm of nightmares, and not the realm of reality. There is at least a chance for us to avert, or at the very least allay, an unthinkable unraveling of pain. We must claim it.

Solace can be found in the solutions we know. We have a treasure trove of traditions, tested techniques, ancient ideas and rich philosophies that can guide us. People have spent millennia understanding our climate, the ecosystems that sustain us and how we can thrive among them. We have a world filled with countless people devoted to dignity. Such is the hope we hold.

■ ■ ■ ■ ■

Life, our procession of memory, is deeply fragile. From the smallest cell to the largest star, all is mortal. Eventually, our nearby Sun will exhaust its fuel. In 100 million years, there will be little evidence left of the human story.

We are destined to oblivion. But as the historian Howard Zinn wrote, "the future is an infinite succession of presents, and to live now as we think human beings should live, in defiance of all that is bad around us, is itself a marvellous victory."[11]

We all have gifts to give, abilities to use, people around us to inspire. We can interlock our passions with activism. Those fortunate enough to live in relatively free societies can put voices, bodies and hearts on the line. The botanist Robin Wall Kimmerer compels us toward the ideal of reciprocity, to acknowledge the gifts we have, that we can give in return for all that has been given to us.

Kimmerer asks:

> What does it mean to be an educated person? It means that you know what your gift is and how to give it on behalf of the land and of the people, just like every single species has its own gift. And if one of those species and the gifts that it carries is missing in biodiversity, the ecosystem is desperate, the ecosystem is too simple. It doesn't work as well when that gift is missing.[12]

Ecology at its heart is love for all beings, a living acknowledgment of all that allows us to be: our solar system, and its unique capacity to shield our planet from threats; the precious swirling rock that is our Earth, and its exceptional atmosphere which allows for the gifts of breath and nourishment; the efforts of ecosystems, fellow beings and ancestors that have brought us here. We exploit what we ignore and devalue. But we live for, we think for, we fight for what we love.[13]

The infinite stakes rob us of patience. Left unchecked, climate violence spells a radical and regressive reconfiguration of our world. But if we unleash a real transformation that can slash emissions, improve public health and offer people more dignified lives, then we can pry open the possibility of achieving a more beautiful and just future.

As the anthropologist Fernando Coronil reminds us:

> The embers of the past and the poetry of the future will continue to conjure up images of worlds free from the horrors of history. Politics will remain a battle of desires waged on an uneven terrain. But as long as people find themselves without a safe and dignified home in the world, utopian dreams will continue to proliferate, energizing struggles to build a world made of many worlds, where people can dream their futures without fear of waking up.[14]

Climate change represents the memory we could be, the possibility of pain or prosperity. It is up to us to evict the certainties of suffering and craft a world that resembles the dignity of its people. Let us weave together the protections of solidarity. Let us suture the separations between culture and nature, between us and the other. Let us confront the poverties of humans, in all their layers. Together, we can build a home worthy of its name.

Acknowledgments

These words would not be here without the contributions of thousands of journalists, historians, scientists, activists, justice-seekers, life-givers farmers and artists, to whose insights I am indebted.

This book would also not be here if not for a series of coincidental fortunes, in the form of people. My gratitude flows to Raoul, for the impetus to try, and for the humility of teaching. To Jamie, for his brazen faith and energy. To Dan, for his trust and guidance, and Chris, for his deft book healing.

To Asad, whose relentless vision for justice powers these pages. To Alcides, for showing me how to think, and why. To Adam, for unknowingly revealing the power of language. To Leandro, for the treasures of poetry. To Jia-Hui, for opening the world of activism. To Laura, for the best questions and the gift of endless curiosity. To Harry, for the courage of irreverence. To Sebas, for roots and reasons to dream. To Ali, for lessons of dignity. To Nathan, for the joys of collaboration, and for privilege of learning. To Igor, for discovery and brotherhood. To Justin, for encouragement when it was much needed.

To Leticia, Patricia, David, Julia, Danilo, Karine, Sam, Ruth, Majandra, Alejandra, Tincho, Norka, Kemo, Linh, Yasmin, Sonya, Martin, Raj, Katie, Benja, Adam, Chess, Aghil, Sergio, from whom I have learned more than they will ever know. To Giorgos, Carme, Mario and Diego, whose pedagogy did much to clarify impulses.

To Checho, to the Valbuena family and to Albeiro, for their unmerited generosity. To all who shared their stories and opened their homes.

To those indispensable companions across distance, Astor, Darío, Eduardo and Svetlana.

To Mitya, my endless inspiration since childhood. To Katya, for the energies of inquisitiveness and persistence. To Sasha, for lessons in joy and unthought kindness. To my grandparents, for the memory they have lent me. To Mamuchka and Papuchka, my dearest parents, for the precious gifts of life, love and learning.

Y para Mar, for endless patience, unwarranted support and for making hope a horizon.

While all these wonderful people and many more have made this possible, the many mistakes and omissions of this book are mine alone.

Notes

Chapter 1: The Might of Memory

1. Rebecca Kessler, "Written in Baleen," *Aeon*, 28 Sep 2016. nin.tl/Whaleshistory
2. Christophe Bonneuil and Jean-Baptiste Fressoz, *The Shock of the Anthropocene*, Verso 2017, p. 5.
3. Food and Agriculture Organisation of the United Nations, *The State of World Fisheries and Aquaculture*, FAO, 2016.
4. T. W. Crowther et al., "Mapping Tree Density at a Global Scale," *Nature*, vol. 525, no. 7568, 2015.
5. Humanity first overshot the Earth's annual regenerative capacity in the early 1970s. That is to say our species reached a point where every year we started to consume more resources and produce more waste than the natural bioproductivity of the planet and ecosystem functions are able to regenerate and safely absorb in a year.
6. *New Internationalist*, 473, June 2014, p. 20.
7. WWF, Living Planet Report, 2016, nin.tl/WWF2016
8. *The New Plastics Economy*, Ellen MacArthur Foundation, nin.tl/PlasticsMacArthur
9. Stuart L. Pimm et al., "The Biodiversity of Species and Their Rates of Extinction, Distribution and Protection," *Science*, vol. 344, no. 6187, 2014. Researchers agree that the rate of extinction of species is many times the pre-human rate. The range of estimates is between 1,000 and 10,000 times. For the quote, see Gerardo Ceballos, Paul R. Ehrlich, and Rodolfo Dirzo, "Biological Annihilation Via the Ongoing Sixth Mass Extinction Signaled by Vertebrate Population Losses and Declines," *Proceedings of the National Academy of Sciences*, vol. 114, no. 30, 2017.
10. World Health Organization, *Ambient Air Pollution: A Global Assessment of Exposure and Burden of Disease*, p. 33.
11. Gardiner Harris, "Holding Your Breath in India," *New York Times*, 29 May 2015.
12. Philip J. Landrigan, et al., "The Lancet Commission on Pollution and Health," *The Lancet*, vol. 391, no. 10119, 2017.

13. Will Steffen, et al., "Planetary Boundaries," *Science*, vol. 347, no. 6223, 2015.
14. *The Economist*, "Weather-Related Disasters are Increasing," 29 Aug 2017.
15. Cited in Reed F. Noss, "Some Principles of Conservation Biology, as They Apply to Environmental Law," *Chi.-Kent L Rev*, vol. 69, 1993.
16. Enrique Leff, *Saber Ambiental: Sustentabilidad, Racionalidad, Complejidad, Poder*, Siglo XXI Editores, 1998.
17. Cited in Leonardo Boff, *Ecologia: grito da terra, grito dos pobres*, Ed Ártica 1995, p. 19.

Chapter 2: Separation

1. Horacio Machado Aráoz, "Orden neocolonial, extractivismo y ecología política de las emociones,"' *Revista Brasileira de Sociología de las Emociones*, vol. 34, 2013.
2. Angela Guimaraes Pereira and Silvio Funtowicz, *Science, Philosophy and Sustainability: The End of the Cartesian Dream*, Routledge, 2015.
3. Jeremy L. Caradonna, *Sustainability: A History*, Oxford University Press, p. 30.
4. David Bohm, *Wholeness and the Implicate Order*, Routledge 1980, p. 3.
5. Val Plumwood, "Ecological Ethics from Rights to Recognition," in *Global Ethics and Environment*, Routledge 2003, p. 194.
6. Barry Commoner, *The Closing Circle*, Knopf Doubleday, 2014.
7. Philippe Vandenkoornhuyse et al., "The Importance of the Micro-biome of the Plant Holobiont," *New Phytologist*, vol. 206, no. 4, 2015.
8. Ed Yong, *I Contain Multitudes*, Bodley Head, 2016.
9. Thomas Berry, *Evening Thoughts*, Sierra Club Books, 2006, p. 17.
10. Alex Carr Johnson, "How to Queer Ecology: One Goose at a Time," *The Environment in Anthropology*, 2016, NYU Press.
11. Timothy Morton, "Guest Column: Queer Ecology," *PMLA*, vol. 125, no. 2, 2010.
12. Vilma Pérez Pino, *Educación Ambiental y Cosmovisión de los Pueblos Originarios*, IECTA, 2004.
13. David George Haskell, nin.tl/keytolife
14. This section draws on the work of countless thinkers, from Manfred Max-Neef to Donella Meadows. A good place to start is: Ilya Prigo-nine and Isabelle Stengers, *The End of Certainty*, Simon and Schuster, 1997; Fritjof Capra and Pier Luigi Luisi, *The Systems View of Life*, Cambridge University Press, 2014.
15. Russell Bonduriansky and Troy Day, *Extended Heredity*, Princeton University Press, 2018, p. 7.
16. Philippe Vandenkoornhuyse, op cit; Frédéric Delsuc, Henner Brink-

mann and Hervé Philippe, "Phylogenomics and the Reconstruction of the Tree of Life," *Nature Reviews Genetics*, vol. 6, no. 5, 2005.

17. Bratislav Mišić and Olaf Sporns, "From Regions to Connections and Networks," *Current Opinion in Neurobiology*, vol. 40, 2016.

18. Russell Bonduriansky and Troy Day, op cit, p. 7.

19. For insights on the charm of precision, see Theodor Porter, *Trust in Numbers*, Princeton University Press, 1995. For broader insights on how recent scientific insights are challenging notions of punishment, see Raoul Martinez, *Creating Freedom*, Canongate Books, 2016.

20. Robert Sapolsky, *Behave: The Biology of Humans at Our Best and Worst*, Penguin, 2017.

21. Mario Giampetro, *Multi-scale Integrated Analysis of Agro-Ecosystems*, CRC Press, 2003.

22. The work of Jerome Ravetz and Silvio Funtowicz is path-breaking in this regard. See, Silvio O. Funtowicz and Jerome R. Ravetz, "The Worth of a Songbird: Ecological Economics as a Post-normal Science," *Ecological Economics*, vol. 10, no. 3, 1994.

23. Carolyn Merchant, *Autonomous Nature*, Routledge 2016, p. 5.

Chapter 3: Origins

1. Madhav Gadgil and Ramachandra Guha, *This Fissured Land*, Oxford University Press, 1993.

2. David Graeber and David Wengrow, "How to Change the Course of Human History," *Eurozine*, 2 Mar 2018.

3. Jason Moore and Raj Patel, *A History of the World in Seven Cheap Things*, University of California Press, p. 2.

4. Stephen J. Pyne, "Frontiers of Fire," in *Ecology and Empire*, Keele University Press, 1997, p. 23.

5. Dan Charles, "From Cattle To Capital," *NPR*, 15 Nov 2017.

6. Ashley Dawson, *Extinction: A Radical History*, OR Books, 2016.

7. Timothy A. Kohler et al., "Greater Post-Neolithic Wealth Disparities in Eurasia than in North America and Mesoamerica," *Nature*, vol. 551, no. 7682, 2017; for important rebuttal, see David Graeber and David Wengrow, "How to Change the Course of Human History," *Eurozine*, 2 Mar 2018.

8. James C. Scott, *Against the Grain*, Yale University Press, 2017.

9. Tim Ingold, *The Perception of the Environment*, Routledge 2000, pp. 73–74.

10. John Lanchester, "The Case Against Civilization," *New Yorker*, 18 Sep 2017.

11. Richard H. Grove, *Green Imperialism*, Cambridge University Press, 1996, p. 18–19.

12. David R. Montgomery, *Dirt: The Erosion of Civilizations*. University of California Press, 2012, p. 51.
13. Jason Moore and Raj Patel, op cit, p. 2.
14. David R. Montgomery, op cit, p. 62.
15. Ashley Dawson, op cit, 2016.
16. Joachim Radkau, *Nature and Power*. Cambridge University Press, 2008, p. 32.
17. See the work of Karl Butzer: Karl W. Butzer, "Collapse, Environment, and Society," *Proceedings of the National Academy of Sciences*, vol. 109, no. 10, 2012.
18. Joachim Radkau, op cit, p. 139.
19. Ibid, p. 96.
20. Ibid, pp. 109–114.
21. Roger D. Masters, *Fortune is a River*, The Free Press, 1998.
22. Tamim Ansary, *Destiny Disrupted*, PublicAffairs, 2009, p. 6.
23. Matthieu Ricard, *A Plea for the Animals*, Shambhala Publications, 2016, p. 200.
24. Ashley Dawson, *Extinction: A Radical History*, OR Books, 2016.

Chapter 4: Colonialism: The Acceleration

1. Cited in Fernando Rovetta Klyver, *El Descubrimiento de los Derechos Humanos*, IEPALA 2008, p. 162.
2. Eduardo Galeano, "Mundo: Cuatro frases que hacen crecer la nariz de Pinocho," *Servindi*, nin.tl/Galeano
3. Cited in Nathan Wolski, "All's Not Quiet on the Western Front—Rethinking Resistance and Frontiers in Aboriginal Historiography," *Colonial Frontiers: Indigenous-European Encounters in Settler Societies*, 2001, p. 230.
4. Horacio Machado Aráoz, "El territorio moderno y la geografía (colonial) del capital," *Memoria y Sociedad*, vol. 19, no. 39, 2015; Vumbi-Yoka Mudimbe, *The Invention of Africa*, Indiana University Press, 1988.
5. Eyal Weizman and Fazal Sheikh (photographs), *Colonization as Climate Change in the Negev Desert*, Steidl, 2015.
6. Susan Crate, "Viliui Sakha of Subartic Russia and Their Struggle for Environmental Justice," *Environmental Justice and Sustainability in the Former Soviet Union*, MIT Press, 2009, p. 190.
7. Kendall W. Brown, *A History of Mining in Latin America*, University of New Mexico Press, 2012.
8. This framing draws on the writings of Danilo Urrea and Tatiana Roa Avendaño on the role of nature in the Colombian peace process. See:

"La cuestión ambiental," *Y sin embargo, se mueve,* Ediciones Antropos, 2015.

9. Richard Drayton, *Nature's Government,* Yale University Press, 2000, p. 229.
10. James Ferguson, *Global Shadows,* Duke University Press, 2006, p. 39.
11. David R. Montgomery, *Dirt: The Erosion of Civilizations,* University of California Press, 2012, p. 110.
12. Michael Perelman, *The Invention of Capitalism,* Duke University Press, 2000, p. 52.
13. Cited in Robert H. Nelson, "Environmental Colonialism: 'Saving' Africa from Africans," *Independent Review,* vol. 8, no. 1, 2003.
14. William M. Adams, "Nature and the Colonial Mind," in William M. Adams and Martin Mulligan, eds, *Decolonizing Nature,* Earthscan, 2003, p. 25.
15. Diana Davis, *Resurrecting the Granary of Rome,* Ohio University Press, 2007, pp. 4–6.
16. Per Lindskog and Benoit Delaite, "Degrading Land," *Environment and History,* vol. 2, no. 3, 1996.
17. Jason W. Moore, "Madeira, Sugar, and the Conquest of Nature in the 'First' Sixteenth Century," *Review (Fernand Braudel Center),* 2009.
18. David R. Montgomery, op cit, p. 231.
19. Corey Ross, "The Plantation Paradigm," *Journal of Global History,* vol. 9, no. 1, 2014, p. 49.
20. Joachim Radkau, *Nature and Power,* Cambridge University Press, 2008, p. 191.
21. John Bellamy Fosters, Brett Clark and Richard York, *The Ecological Rift,* NYU Press, 2011, p. 364.
22. David R. Montgomery, op cit, p. 187.
23. Ibid, p. 188.
24. A. Gioda, C. Serrano and A. Forenza, "Dam Collapses in the World," *La Houille Blanche,* vol. 4, 2002; Jason W. Moore, "Silver, Ecology and the Origins of the Modern World, 1450–1640," *Rethinking Environmental History,* Altamira Press, 2007, p. 132.
25. Clive Ponting, *A New Green History of the World,* Random House, 2007, p. 189.
26. William M. Adams, op cit, p. 30.
27. Peter Boomgaard, "Environmental Impact of the European Presence in Southeast Asia, 17th-19th Centuries," *Illes i imperis,* vol. 1 1998, p. 28.
28. Lucy Jarosz, "Defining and Explaining Tropical Deforestation," *Economic Geography,* vol. 69, no. 4, 1993.

29. John F. Richards, *The Unending Frontier*, University of California Press, 2003, p. 463.
30. Callum Roberts, *The Unnatural History of the Sea*, Island Press, 2007.
31. Andrew Stuhl, *Unfreezing the Arctic*, University of Chicago Press, 2016, p. 7.
32. Rafael Bernal, *Mestizaje y criollismo en la literatura de la Nueva España del siglo XVI*, Fondo de Cultura Económica, 2015, pp. 156–159.
33. Aimé Césaire, *Discourse on Colonialism*, Monthly Review Press, 2000, p. 42.
34. Andrés Reséndez, *The Other Slavery*, Houghton Mifflin, 2016, p. 14.
35. Stephen Mosley, *The Environment in World History*, Routledge, 2010.
36. Jason W. Moore, "The Capitalocene, Part I," *The Journal of Peasant Studies*, vol. 44, no. 3, 2017.
37. José Manuel de Prada-Samper, "The Forgotten Killing Fields," *Historia*, vol. 57, 2012.
38. Mohamed Adhikari, *The Anatomy of a South African Genocide*, Ohio University Press, 2011, p. 17.
39. Calla Wahlquist, "Map of Massacres of Indigenous People Reveals Untold History of Australia, Painted in Blood," *Guardian*, 5 July 2017.
40. University of Newcastle, Australia, "Colonial Frontier Massacres in Eastern Australia, 1788–1872," nin.tl/Ausmassacres
41. Pankaj Mishra, "How Colonial Violence Came Home," *Guardian*, 10 Nov 2017.
42. Timothy Snyder, *Black Earth*, Bodley Head, 2015.
43. Luciano Andrés Valencia, "El Genocidio Fueguino," *Resumen—Chile*, 25 Nov 2016.
44. Adolfo Colombres, *La colonización cultural de la América Indígena*, Ediciones del Sol, 1987, pp. 82–83.
45. Daniel Wermus, *¡Madre Tierra! Por el renacimiento indígena*, Ediciones Abya Yala, pp. 189–200.
46. Laszlo Makra and Peter Brimblecombe, "Selections from the History of Environmental Pollution, *International Journal of Environment and Pollution*, vol. 22, no. 6, 2004.
47. Herbert S. Klein and Francisco Vidal Luna, *Slavery in Brazil*, Cambridge University Press, 2009, p. 6.
48. Andrés Reséndez, op cit, p. 26.
49. Andrés Reséndez, op cit, p. 25.
50. Ibid, p. 28.
51. Beatriz Scharrer Tamm, *Azúcar y trabajo*, CIESAS, 1997, p. 150.
52. Andrés Reséndez, op cit.
53. Marcus Rediker, *The Slave Ship: A Human History*, Viking, 2007.
54. Walter Johnson, *River of Dark Dreams*, cited in Daniel Hartley,

"Anthropocene, Capitalocene, and the Problem of Culture," *Anthropocene or Capitalocene?* PM Press, 2016.

55. Andrés Reséndez, op cit, p. 112.
56. Nicholas A. Robins, *Mercury, Mining, and Empire*, Indiana University Press, 2011.
57. Pat Hudson, "Slavery, the Slave Trade, and Economic Growth: A Contribution to the Debate," *Emancipation and the Remaking of the British Imperial World*, Manchester University Press, 2014.
58. Clare Anderson, "Legacies of a British Penal Colony," nin.tl/Andamanblog
59. Corey Ross, *Ecology and Power in the Age of Empire*, Oxford University Press, 2017, p. 101.
60. Adam Hochschild, *King Leopold's Ghost*, Houghton Mifflin Harcourt, 1999.
61. Bronwen Everill, "Demarginalizing West Africa in the Age of Revolutions," nin.tl/Demarginalizing
62. Adam Hochschild, op cit.
63. J. R. McNeill, *Mosquito Empires*, Cambridge University Press, 2010, p. 16.
64. Justin McBrian, "Accumulating Extinction," *Anthropocene or Capitalocene?* PM Press, 2016, pp. 116–137.
65. Massimo Livi Bacci, *Conquest: The Destruction of the American Indios*, Polity, 2008.
66. Cited in Jason Hickel, "Enough of Aid—Let's Talk Reparations," *Guardian*, 27 Nov 2015.
67. Adam Hochschild, op cit.
68. Jason Hickel, *The Divide*, William Heinemann, 2017.
69. Joseph Inikori, *Africans and the Industrial Revolution in England*, Cambridge University Press, 2002.
70. Angus Maddison, *The World Economy*, OECD, 2006.
71. Mike Davis, "The Origin of the Third World," *Antipode*, vol. 32, no. 1, 2000.
72. David R. Montgomery, op cit, p. 100.
73. John L. Brooke, *Climate Change and the Course of Global History*, Cambridge University Press 2014, p. 496.
74. Jason Hickel, *The Divide*, Heinemann, 2017
75. Jason W. Moore, "The Capitalocene," op cit.
76. Joachim Radkau, op cit, p. 111.
77. Jeremy L. Caradonna, *Sustainability: A History*, Oxford University Press, p. 33.
78. Jason W. Moore, "Amsterdam Is Standing on Norway," *Journal of Agrarian Change*, vol. 10, no. 1, 2010.

79. Joachim Radkau, op cit, p. 173

80. Ibid, p. 117.

81. Satpal Sangwan, "From Natural History to History of Nature," *Les Sciences hors d'Occident au 20ème siècle*, Orstom, 1996.

82. This period is partly known as the Second Serfdom. See: Eric R. Wolf, *Pathways of Power*, University of California Press, 2001, pp. 272–288; Richard Drayton and David Motadel, "Discussion: The Futures of Global History," *Journal of Global History*, vol. 13, no. 1, 2018.

83. Jason W. Moore, "Dutch Capitalism and Europe's Great Frontier," in *The Longue Durée and World-Systems Analysis*, SUNY Press, 2012, p. 66.

84. Ibid.

85. Veli-Pekka Lehtola, "Sámi Histories, Colonialism, and Finland," *Arctic Anthropology*, vol. 52, no. 2, 2015.

86. Michael Perelman, op cit.

87. Joachim Radkau, op cit, p. 192.

88. Ibid, p. 182

89. Michael Perelman, op cit, p.14

90. Jason W. Moore, "The Capitalocene," op cit.

91. Joachim Radkau, op cit., p. 192.

92. Jason Hickel, op cit, pp. 78–79.

93. Liz Alden Wily, "Looking Back to See Forward," *Journal of Peasant Studies*, vol. 39, no. 3-4, 2012.

94. Alexander Etkind, *Internal Colonization: Russia's Imperial Experience*, Polity, 2011.

95. Walter Richmond, *The Circassian Genocide*, Rutgers University Press, 2013.

96. Paul Josephson, *Industrialized Nature*, Island Press, 2002, p. 178.

97. Minako Sakata, "Japan: Incorporation of Hokkaido," nin.tl/Hokkaido.

98. Nic Paget-Clarke, "Interview with Dr Tewolde BG Egziabher," *In Motion Magazine*, 15 Dec 2002.

99. Dasha Shkurpela, "Dacha," *The Calvert Journal*, 2017.

100. Andreas Malm, "Who Lit This Fire?," *Critical Historical Studies*, vol. 3, no. 2, 2016.

101. Robert B. Marks, *China: An Environmental History*, Rowman & Littlefield, 2017, p. 315.

102. Sam Kean, "The Soviet Era's Deadliest Scientist Is Regaining Popularity in Russia," *Atlantic*, 19 Dec 2017.

103. Peter Pringle, *Murder of Nikolai Vavilov*, JR Books, 2009.

104. Lynne Viola, *The Unknown Gulag*, Oxford University Press, 2007, p. 4.

105. Jeffrey S. Hardy, *The Gulag After Stalin*, Cornell University Press, 2016, p. 16.
106. Peter Radetsky, *The Soviet Image*, Chronicle Books, 2007, p. 88.
107. Golfo Alexopoulos, *Illness and Inhumanity in Stalin's Gulag*, Yale University Press, 2017.
108. Michael Kort, *The Soviet Colossus: History and Aftermath*, Routledge 2015, p. 217.
109. Ibid.
110. Zhores A. Medvedev and Roy A. Medvedev, *The Unknown Stalin*, IB Tauris, 2006, pp. 152–170.
111. Anne Applebaum, *Gulag: A History of the Soviet Camps*, Penguin, 2012.
112. Known as the *Holofdomor* (Ukraine) and *Zhasandy Asharshylyk* (Kazakhstan).
113. Cited in Paul Josephson, op cit, p. 188.
114. Ian D. Whyte, *A Dictionary of Environmental History*, IB Tauris, 2013, p. 39.
115. Laura A. Henry, "Thinking Globally, Limited Locally," *Environmental Justice and Sustainability in the Former Soviet Union*, MIT Press, 2009, p. 49.
116. Olivia J. Wilson and Geoff A. Wilson, "East Germany," *Environmental Problems in East-Central Europe*, Routledge, 2002, p. 141.
117. Claudio O. Deland and Zhen Yuan, *China's Grain for Green Program*, Springer, 2014, p. 8.
118. Ibid.
119. Joachim Radkau, op cit, page 99.
120. Philip Ball, *The Water Kingdom*, University of Chicago Press, 2017.
121. For context, see: Angel Hsu and William Miao, "28,000 Rivers Disappeared in China," *The Atlantic*, 29 Apr 2013.
122. Celestino Andrés Araúz, "Un sueño de siglos: El Canal de Panamá," *Revista Debate*, no. 21, Sep 2013.
123. Margit Szöllösi-Janze, "The Scientist as Expert," *One Hundred Years of Chemical Warfare*, Springer, 2017, pp. 11–23.
124. Christophe Bonneuil and Jean-Baptiste Fressoz, *The Shock of the Anthropocene*, Verso, 2017; Edmund Russell, *War and Nature*, Cambridge University Press, 2001.
125. Daniel Yergin, *The Prize*, Simon & Schuster, 1991, p. 170.
126. Christophe Bonneuil and Jean-Baptiste Fressoz, *The Shock of the Anthropocene*, Verso, 2017.
127. Margit Szöllösi-Janze, op cit.
128. Ibid.
129. Michael J. Watts, *Silent Violence*, University of Georgia Press, 2013.

130. Silke Stroh, "Towards a Postcolonial Environment?," *Local Natures, Global Responsibilities,* Rodopi, 2010, p. 197.

131. Clive Ponting, op cit, p. 192.

132. John H. Bodley, *Anthropology and Contemporary Human Problems,* Rowman Altamira, 2012, p. 47.

133. Such as in the case of the Rio Negro massacres in Guatemala.

134. Autumn Spanne, "Why is Honduras the World's Deadliest Country for Environmentalists?," *Guardian,* 7 Apr 2016.

135. Darío Aranda, "Qué hay detrás de la campaña antimapuche," *La Vaca,* 27 Nov 2017.

136. Gestión, "Concesiones mineras ocupan la quinta parte del territorio del Perú," 14 Sep 2014, nin.tl/Peru

Chapter 5: Fossil Fuels, Furious Flames

1. Daniel Yergin, *The Prize,* Simon & Schuster, 1991, p. 73.

2. Leonardo Maugeri, *The Age of Oil,* Praeger, 2006, p. 3.

3. Daniel Yergin, op cit, p. 135.

4. Laszlo Makra and Peter Brimblecombe, "Selections from the History of Environmental Pollution," *International Journal of Environment and Pollution,* vol. 22, no. 6, 2004.

5. James Marriott and Mika Minio-Paluello, *The Oil Road,* Verso, 2013.

6. Andreas Malm, "Who Lit This Fire?," *Critical Historical Studies,* vol. 3, no. 2, 2016.

7. Joe Romm, "Humans Boosting CO_2 14,000 Times Faster than Nature, Overwhelming Slow Negative Feedbacks," *Think Progress,* 28 Apr 2008.

8. Andreas Malm, op cit.

9. Ibid.

10. See: Douglas A. Blackmon, *Slavery By Another Name,* Anchor, 2009.

11. Ihediwa Nkemjika Chimee, "Coal and British Colonialism in Nigeria," *RCC Perspectives* 5, 2014, pp. 19–21.

12. Tim Maly, "A Brief History of Human Energy Use," *The Atlantic,* 13 Nov 2015. See also Vaclav Smil, *Energy Transitions,* ABC-CLIO, 2010.

13. For an interesting discussion of coal's relevance in history, see: Kenneth Pomweranz, "Without Coal? Colonies? Calculus?," *Unmaking the West,* University of Michigan Press, 2006.

14. Daniel Yergin et al., "Energy Vision 2013," *World Economic Forum,* 2013.

15. Andreas Malm, *Fossil Capital,* Verso, 2016.

16. Jean-François Mouhot, "Past Connections and Present Similarities in Slave Ownership and Fossil Fuel Usage," *Climatic Change,* vol. 105, nos. 1-2, 2011.

17. Imre Szeman, "On the Energy Humanities," *European Union Centre of Excellence Working Papers University of Alberta*, vol. 1, 2015.
18. Stephen J. Pyne, "Frontiers of Fire," *Ecology and Empire*, University of Washington Press, 1997, p. 31.
19. Steffen Böhm, "The Paris Climate Talks and Other Events of Carbon Fetishism," *Verso*, 3 Dec 2015.
20. Charles Perrow, *Organizing America*, Princeton University Press, 2002; Patrick Robbins, "Interrupting the Future," *The Leap Blog*, 10 Apr 2015.
21. Dipesh Chakrabarty, "The Climate of History: Four Theses," *Critical Inquiry*, vol. 35, no. 2, 2009.
22. Hans A. Bear, *Global Capitalism and Climate Change*, Rowman Altamira, 2012, p. 2.
23. Michael Watts, "Petro-violence," *Violent Environments*, Cornell University Press, 2001.
24. Yergin, op cit.
25. Yergin, op cit, p. 97.
26. Patrick Robbins, op cit.
27. Yergin, op cit, p. 236 and p. 55.
28. Ibid, p. 239.
29. Ibid, p. 115.
30. Tatiana Roa Avendaño, "Petróleo y deuda ecológica," *CENSAT Agua Viva*, 2002.
31. Ibid.
32. Eduardo Sáenz Rovner, "La Industria Petrolera en Colombia," *Revista Credencial Historia*, vol. 49, 1994.
33. Tatiana Roa Avendaño, op cit.
34. Organización Nacional Indígena de Colombia, "Sobre el petróleo, algunas preguntas—intervención," Medio Ambiente y Minas y Energía, 29 May 1997.
35. Ibid.
36. Nnimmo Bassey, "Oil and Gas in Africa" in Rowil Aguillon et al., *No More Looting and Destruction!*, Ediciones Abya Yala, 2003, p. 86.
37. Nick Mathiason, "Amnesty Attacks Oil Industry for Decades of Damage in Niger Delta," *Guardian*, 30 Jun 2009; Michael Watts, "Petro-violence," *Violent Environments*, Cornell University Press, 2001.
38. Amnesty International, "Nigeria: A Criminal Enterprise?," 2017.
39. Hannah Summers, "Amnesty Seeks Criminal Inquiry into Shell Over Alleged Complicity in Murder and Torture in Nigeria," *Guardian*, 28 Nov 2017.
40. Michael Watts, op cit.
41. See John Bacher, *Petrotyranny*, Dundurn, 2000.

42. Ramachandra Guha and Joan Martínez Alier, *Varieties of Environmentalism*, Earthscan, 1997, pp. 43–44.
43. Marjorie Housepian Dobkin, "What Genocide? What Holocaust?," *The Armenian Genocide in Perspective*, Transaction Books, 1986, p. 104.
44. Rómulo Betancourt, *Venezuela, Política y Petróleo*, Universidad Católica Andrés Bello, 2007, p. 40.
45. Cornelius Goslinga, "Oil Comes to the Curaçao Islands," *A Short History of the Netherlands Antilles and Suriname*, Springer, 1979, pp. 141–143.
46. Daniel Yergin, op cit, p. 236.
47. Ibid, p. 13.
48. Cited in Steve A. Yetiv, *The Petroleum Triangle*, Cornell University Press, 2011, p. 25.
49. Yergin, p. 393.
50. Ibid, p. 499.
51. Cited in Rob Nixon, *Slow Violence and the Environmentalism of the Poor*, Harvard University Press, 2011, p. 72.
52. Hardy Calvert, "Ken Silverstein's The Secret World of Oil," *Harper's*, 7 Jul 2014.
53. Michael J. Watts, *Silent Violence*, University of Georgia Press, 2013, p. xiv.
54. Imre Szeman and Petrocultures Research Group, *After Oil*, West Virginia University Press, 2016, pp. 29–40.
55. This insight, made by economist Paul Sweezy, is cited in: Dave Holmes, Terry Townsend and John Bellamy Foster, *Change the System, Not the Climate!*, Resistance Books, 2007, pp. 19–20.
56. Imre Szeman, "From Petrocultures to Other Cultures," *The Quad*, 26 Apr 2016.
57. Shannon Hall, "Exxon Knew about Climate Change Almost 40 Years Ago," *Scientific American*, 26 Oct 2015.
58. Geoffrey Supran and Naomi Oreskes, "Assessing ExxonMobil's Climate Change Communications (1977–2014)," *Environmental Research Letters*, vol. 12, no. 8, 2017; James Lawrence Powell, *The Inquisition of Climate Science*, Columbia University Press, 2011; Robert O. Schneider, *When Science and Politics Collide*, ABC-CLIO, 2018, p. 70.
59. Jane Mayer, *Dark Money*, Anchor, 2016.
60. David Coady et al., "How Large Are Global Energy Subsidies?," Working Paper no. 15-105, International Monetary Fund, 2015.
61. Roger Fouquet, "Energy: Muscle, Steam and Combustion," *Nature*, vol. 544, no. 7648, 2017.
62. Cited in Tim Flannery, *Atmosphere of Hope*, Penguin, 2015.

Chapter 6: Human Nature or Human Ignorance?

1. Cited in Mauricio Archila and Martha Cecilia Garica, "Sobre interculturalidad y nociones de desarrollo," *Hasta Cuando Soñemos*, CINEP/Programa por la Paz, 2015, p. 147.
2. Raymond Pierotti, *Indigenous Knowledge, Ecology, and Evolutionary Biology*, Routledge, 2010.
3. Alex Carr Johnson, "How to Queer Ecology: One Goose at a Time," *The Environment in Anthropology*, 2016, NYU Press, p. 313.
4. *No habrá justicia global, hasta que tengamos justicia cognitiva.*
5. Horacio Machado Aráoz, "Orden neocolonial, extractivismo y ecología política de las emociones," *Revista Brasileira de Sociología de las Emociones*, vol. 34, 2013, p. 36.
6. Víctor M. Toledo and Narciso Barrera-Bassols, *La Memoria Biocultural*, Icaria Editorial, 2008, p. 59.
7. Robert Pogue Harrison, *Forests: The Shadow of Civilization*, University of Chicago Press, 2009.
8. Kath Weston, *Animate Planet*, Duke University Press, 2017, p. 8.
9. Tim Ingold, *The Perception of the Environment*, Routledge 2000, p. 43.
10. Bayo Akomolafe, "Free Will Versus Determinism," 21 May 2017, nin.tl/Bayo
11. David R. Montgomery, *Dirt*, University of California Press, 2012, p. 27.
12. Lorne Leslie Neil Evernden, *The Social Creation of Nature*, JHU Press, 1992.
13. Insight from Mandy Haggith, cited in: Ma Tianjie, "On Trees, Poetry and Humankind's Relationship with Nature," *China Dialogue*, 27 Dec 2017.
14. Eduardo Kohn, *How Forests Think*, University of California Press, 2013; Focus on the Global South, *From Latin America to Asia, Learning from our Roots: A Conversation on Vivir Bien*, Aug 2013; Krista Langlois, "When Whales and Humans Talk," *Hakai*, 3 Apr 2018.
15. See Víctor M. Toledo and Narciso Barrera-Bassols, op cit; César Carrillo Trueba, *Pluriverso*, Ediciones Abya-Yala, 2008.
16. Andrew J. Trant et al., "Intertidal Resource Use Over Millennia Enhances Forest Productivity," *Nature communications*, vol. 7, 2016.
17. Vimbai C. Kwashirai, "Environmental History of Africa," *Eolss Unesco Publication*, 2012.
18. Joachim Radkau, *Nature and Power*, Cambridge University Press, 2008, p. 22.
19. Víctor M. Toledo and Narciso Barrera-Bassols, op cit.
20. The World Bank estimates that 80 percent of the planet's biodiversity lies within territories and lands of indigenous peoples.

21. Cited in Stanley Stevens, "The Legacy of Yellowstone," in *Conservation Through Cultural Survival*, Island Press, 1997, p. 27.
22. Ibid.
23. Jonathan Loh and David Harmon, *Biocultural Diversity*, WWF Netherlands, 2014.
24. Guy D. Middleton, "Do Civilisations Collapse?" *Aeon*, 16 Nov 2017.
25. Guy D. Middleton, op cit; G. S. Cumming and G. D. Peterson, "Unifying Research on Social-Ecological Resilience and Collapse," *Trends in Ecology & Evolution* 32(9), 2017.
26. Terry Hunt and Carl Lipo, *The Statues That Walked*, Simon and Schuster, 2011.
27. Guy D. Middleton, op cit.
28. Joseph Tainter, *The Collapse of Complex Societies*, Cambridge University Press, 1988.
29. Samuel Arbesman, *Overcomplicated*, Penguin, 2017.
30. Danny Hillis, "The Enlightenment is Dead, Long Live the Entanglement," *Journal of Design and Science*, 2016; Ian Goldin and Mike Mariathasan, *The Butterfly Defect*, Princeton University Press, 2014.
31. Fikret Berkes, "Traditional Ecological Knowledge in Perspective," in Julian T. Inglis (ed), *Traditional Ecological Knowledge*, International Program on Traditional Ecological Knowledge, 1993, p. 4; Gleb Raygorodetsky, *The Archipelago of Hope*, Pegasus Books, 2017.
32. See *Traditional Ecological Knowledge*, op cit.
33. Chief Robert Wavey, "International Workshop on Indigenous Knowledge and Community Based Resource Management," *Traditional Ecological Knowledge*, op cit, p. 13.
34. David Singh, "The Wave That Eats People—The Value of Indigenous Knowledge for Disaster Risk Reduction," UNISDR, 9 Aug 2011.
35. There is ample ethno-botanical work in this field. See the work of Ramón Mariaca Méndez.
36. Dawit Solomon et al., "Indigenous African Soil Enrichment as a Climate-Smart Sustainable Agriculture Alternative," *Frontiers in Ecology and the Environment*, vol. 14, no. 2, 2016.
37. Coffi Y. Prudencio, "Ring Management of Soils and Crops in the West African Semi-Arid Tropics," *Agriculture, Ecosystems & Environment*, vol. 47, no. 3, 1993.
38. Gerald G. Marten, *Traditional Agriculture in Southeast Asia*, Westview Press, 1986.
39. Sanoussi Atta et al., "Adapting to Climate Variability and Change in Smallholder Farming Communities," *Journal of Agricultural Extension and Rural Development*, vol. 7, no. 1, 2015; Devjyot Ghoshal, "An Ancient Technology Is Helping India's 'Water Man' Save Thousands

of Parched Villages," *Quartz*, 23 March 2015; Mi Zhou and Dorien Braam, "Integrating Resilience in South Asia," *Forced Migration Review*, no. 49, May 2015, p. 64.

40. Víctor M. Toledo and Narciso Barrera-Bassols, op cit, p. 22.
41. George Nicholas, "It's Taken Thousands of Years, But Western Science Is Finally Catching Up to Traditional Knowledge," *The Conversation*, 15 Feb 2018.
42. Philip Ball, *The Water Kingdom*, University of Chicago Press, 2017; Joachim Radkau, op cit, p. 107.
43. Joachim Radkau, op cit, p. 87.
44. David R. Montgomery, op cit, p. 65.
45. Ibid, p. 62
46. Joachim Radkau, op cit, p. 17.
47. José Rojas López, "Una apreciación crítica del modelo trizonal de Humboldt-Codazzi en la geografía de Venezuela," *Procesos Históricos*, vol. 12, 2018; Jeremy L. Caradonna, *Sustainability: A History*, Oxford University Press, p. 50.
48. Naomi Klein, "Dancing the World into Being," *Yes! Magazine*, 5 March 2015.
49. In ancient Greece, Xenophon and Lucretius observed that lead mines in Attica were harmful to human health; Laszlo Makra and Peter Brimblecombe, "Selections from the History of Environmental Pollution," *International Journal of Environment and Pollution*, vol. 22, no. 6, 2004.
50. Joachim Radkau, op cit, pp. 241 and 267
51. Laszlo Makra and Peter Brimblecombe, op cit.
52. Joachim Radkau, op cit, p. 150.
53. Ibid, p. 151.
54. Ibid, p. 173.
55. Richard H. Grove, *Green Imperialism*, Cambridge University Press, 1996, p. 19.
56. Ibid, p. 23.
57. Joachim Radkau, op cit, pp. 119–210.
58. Pankaj Jain, *Dharma and ecology of Hindu communities*, Ashgate, 2011, p. 51.
59. Joachim Radkau, op cit, p. 107.
60. Ibid, p. 143.
61. Ibid, p. 240.
62. Ibid, p. 202.
63. Masao Takenaka, *God is Rice*, Wipf and Stock Publishers, 2009, p. 55.
64. Richard H. Grove, "Colonial Conservation, Ecological Hegemony and Popular Resistance," *Imperialism and the Natural World*, 1990;

Christopher V. Hill, *South Asia: An Environmental History*, ABC-CLIO, 2008, p. 101.

65. Mike Davis, "The Coming Desert," *New Left Review*, vol. 97, Jan–Feb 2016.

66. Joachim Radkau, op cit, p. 113.

67. Richard H. Grove, op cit, p. 30.

68. Ibid.

69. Andrea Wulf, "Alexander von Humboldt," *LA Times*, 5 Jul 2015.

70. Kalyanakrishnan Sivaramakrishnan, "Science, Environment and Empire History," *Environment and History*, vol. 14, no. 1, 2008; Mike Davis, op cit.

71. Deborah Coen, "Climate Change and the Humanities," *Items*, 9 Jan 2018.

72. James Rodger Fleming, *Historical Perspectives on Climate Change*, Oxford University Press 1998, p. 57.

73. Kent A. Klitgaard, "Hydrocarbons and the Illusion of Sustainability," *Monthly Review*, vol. 68, no. 3, Jul 2016.

74. Christophe Bonneuil and Jean-Baptiste Fressoz, *The Shock of the Anthropocene*, Verso, 2016, p. 4.

75. Mike Davis, op cit.

76. Benjamin Franta, "On Its 100th Birthday in 1959, Edward Teller Warned the Oil Industry About Global Warming," *Guardian*, 1 Jan 2018.

77. Yancey Orr, J. Stephen Lansing, and Michale R. Dove, "Environmental Anthropology," *Annual Review of Anthropology*, vol. 44, no. 18, 2015.

78. Joseph Romm, *Climate Change: What Everyone Needs to Know*, Oxford University Press, 2016, p. xv.

79. Philip Shabecoff, "Global Warming Has Begun, Expert Tells Senate," *New York Times*, 1988.

80. "1992 World Scientists' Warning to Humanity," nin.tl/Warning

81. Kevin Anderson, "Enthusiasm Over Small Fall in EU Emissions Masks Underlying Apathy on 2°C," nin.tl/Anderson

82. Arnim Scheidel et al., "Ecological Distribution Conflicts as Forces for Sustainability: An Overview and Conceptual Framework," *Sustainability Science*, 2017.

83. Joachim Radkau, op cit, pp. 272, 301.

84. Steve Rayner, "Uncomfortable Knowledge," *Economy and Society*, vol. 41, no. 1, 2012.

85. David P. Barash, "Paradigms Lost," *Aeon*, 27 Oct 2015.

86. Thomas S. Kuhn, *The Structure of Scientific Revolutions*, University of Chicago Press, 2012.

87. Cited in Tim Ingold, *The Perception of the Environment*, Routledge 2000, p. 55.

88. Tim Ingold, op cit, Routledge, 2000.
89. Robert Epstein, "The Empty Brain," *Aeon*, 18 May 2016.
90. Robin Wall Kimmerer, *Braiding Sweetgrass*, Milkweed Editions, 2013.
91. The work of Carlos Lenkersdorf is emblematic in this regard. See: *Aprender a Escuchar*, Plaza y Valdés, 2008; *Conceptos tojolabales de filosofía y del altermundo*. Plaza y Valdés, 2004.
92. Peter Cryle and Elizabeth Stephens, *Normality: A Critical Genealogy*, University of Chicago Press, 2017.
93. Benjamin Zajicek, *Scientific Psychiatry in Stalin's Soviet Union*, University of Chicago Press, 2009.
94. Angela Saini, *Inferior*, Beacon Press, 2017; Cordelia Fine, *Delusions of Gender*, WW Norton, 2010.
95. Robert L. Berger, "Nazi Science—The Dachau Hypothermia Experiments," *New England Journal of Medicine*, vol. 322, no. 20, 1990; Allen M. Hornblum, *Acres of Skin*, Routledge, 2013; Susan M. Reverby, *Examining Tuskegee*, University of North Carolina Press, 2009; Harriet A. Washington, *Medical Apartheid*, Doubleday Books, 2006.
96. Rohan Deb Roy, "It's Time to Decolonize Science and End Another Imperial Era," *Independent*, 13 Apr 2018.
97. Linda Tuhiwai Smith, *Decolonizing Methodologies*, Zed, 2013.
98. Libby Robin, "Ecology: A Science of Empire?," *Ecology and Empire*, University of Washington Press, 1997.
99. Ibid.
100. Andrea Saltelli and Silvio Funtowicz, "What Is Science's Crisis Really About?," *Futures*, vol. 91, Aug 2017.
101. Sabina Siebert, Laura M. Machesky and Robert H. Insall, "Overflow in Science and Its Implications for Trust," *Elife*, vol. 14, no. 4, 2015.
102. Elijah Millgram, *The Great Endarkenment*, Oxford University Press, 2015.
103. See Philip Mirowski, *Science-mart*, Harvard University Press, 2011.
104. Joshua J. Tewksbury et al., "Natural History's Place in Science and Society," *BioScience*, vol. 64, no. 4, 2014; Ismael Ràfols and Jack Stilgoe, "Who Benefits from Biomedical Science?," *Guardian*, 16 Mar 2018.
105. Andrea Saltelli and Silvio Funtowicz, "What is Science's Crisis Really About?," *Futures*, vol. 91, Aug 2017.
106. R. Boden and D. Epstein, "Managing the Research Imagination?," *Globalisation, Societies and Education*, 2006, pp. 223–236.
107. Cristin E. Kearns, Laura A. Schmidt and Stanton A. Glantz, "Sugar Industry and Coronary Heart Disease Research," *JAMA Internal Medicine*, vol. 176, no. 11, 2016.
108. Cited in Andrea Saltelli, "Ethics of Quantification," presented 26 Feb 2016, nin.tl/Saltelli

109. Naomi Oreskes and Erik M. Conway, *Merchants of Doubt*, Bloomsbury, US, 2011; Steve Coll, *Private Empire*, Penguin, 2012.
110. Robert J. Brulle, "Institutionalizing Delay," *Climatic Change*, vol. 122, no. 4, 2014.
111. John PA Ioannidis, "Why Most Published Research Findings Are False," *PLoS medicine*, vol. 2, no. 8, 2005.
112. Richard Horton, "Offline," *Lancet*, vol. 385, no. 9976, 2015.
113. Leonard P. Freedman, Iain M. Cockburn and Timothy S. Simcoe, "The Economics of Reproducibility in Preclinical Research," *PLoS biology*, vol. 13, no. 6, 2015; C. Glenn Begley and John PA Ioannidis, "Reproducibility in Science," *Circulation Research*, vol. 116, no. 1, 2015.
114. Daniel Engber, "Science Is Broken," *Slate*, 5 May 2017.
115. Retraction Watch, "They Thought They Might Solve the World's Energy Problems," 5 Jan 2018, nin.tl/Retraction
116. Monya Baker, "1,500 Scientists Lift the Lid on Reproducibility," *Nature*, 25 May 2016.
117. Iain Chalmers and Paul Glasziou, "Avoidable Waste in the Production and Reporting of Research Evidence," *Obstetrics & Gynecology*, vol. 114, no. 6, 2009.
118. See the work of Andrea Saltelli.
119. Michael Schulson, "A Remedy for Broken Science, Or an Attempt to Undercut It?," *Undark*, 18 Apr 2018.
120. Alvin M. Weinberg, "Science and Trans-science," *Minerva*, vol. 10, no. 2, 1972.
121. Silvio O. Funtowicz and Jerome R. Ravetz, "Uncertainty, Complexity and Post-normal Science," *Environmental Toxicology and Chemistry*, vol. 3, no. 12, 1994.
122. Johannes Mengel, "Q&A with Daniel Sarewitz: What Is the Role of Science in a Post-normal World?" *Medium*, 27 Feb 2017.
123. Personal interview of Maria Faciolince with Danilo Urrea.
124. Silvio O. Funtowicz and Jerome R. Ravetz, op cit, pp. 146–161.
125. Gabriel Popkin, "How Scientists And Indigenous Groups Can Team Up to Protect Forests and Climate," Smithsonian, 3 May 2017; Jerome Ravetz, "Democratising Science in an Age of Uncertainty," *Great Transition*, June 2016.
126. Masanobu Fukuoka, *The One-Straw Revolution*, New York Review of Books, 2009, p. 125.
127. Darío Aranda, "Andrés Carrasco, científico y militante: gracias," *La Vaca*, 10 May 2015.
128. As Robin Kimmerer writes, "Knowledge should be coupled with the responsibility to use that knowledge wisely.'

129. Maarten Boudry and Massimo Pigliucci, eds, *Science Unlimited?*, University of Chicago Press, 2018.

130. Zoe Todd, "An Indigenous Feminist's Take on the Ontological Turn," *Journal of Historical Sociology*, vol. 29, no. 1, 2016.

131. Víctor M. Toledo and Narciso Barrera-Bassols, op cit, p. 65.

132. Boaventura de Sousa Santos, "Más allá del pensamiento abismal," in *Pluralismo epistemológico*, Muela del Diabo Editores, 2009.

133. Rodolfo Stavenhagen, *Los pueblos originarios*, CLACSO, Buenos Aires, 2010, p. 14.

134. Gregory Cajete, *Look to the Mountain*, Kivaki Press, 1994; Robin Wall Kimmerer, "Weaving Traditional Ecological Knowledge Into Biological Education," *AIBS Bulletin*, vol. 52, no. 5, 2002.

135. Leath Tonino, "Two Ways of Knowing," *The Sun Magazine*, Apr 2016.

136. Susanna B. Hecht and Darrell A. Posey, "Preliminary Results on Soil Management Techniques of the Kayapó Indians," *Advances in Economic Botany*, vol. 7, 1989.

137. Víctor M. Toledo and Narciso Barrera-Bassols, op cit.

138. Ibid, p. 92.

139. Richard J. Evans, *The Pursuit of Power*, Penguin 2016, p. 189.

140. Patrick D. Nunn and Nicholas J. Reid, "Aboriginal Memories of Inundation of the Australian Coast Dating from More than 7000 Years Ago," *Australian Geographer*, vol. 47, no. 1, 2016.

141. Robin Wall Kimmerer, "Weaving...," op cit. See also the work of Inuit Qaujisarvingat.

142. Benjamin S. Orlove, John C. H. Chiang and Mark A. Cane, "Ethnoclimatology in the Andes," *American Scientist*, vol. 90, no. 5, 2002.

143. William Barr, "Inuit Folklore Kept Alive Story of Missing Franklin Expedition to North-West Passage," *The Conversation*, 19 Sep 2014.

144. Gleb Raygorodetsky, *The Archipelago of Hope*, Pegasus Books, 2017; Zoe Todd, op cit.

145. Kieran Guilbert, "Centuries-old African Soil Technique Could Combat Climate Change: Scientists," *Reuters*, 21 Jun 2016.

146. Monica di Donato, "Etnoecología, La memoria de la especie humana," *Minerva*, Oct 2009.

147. Jeremy L. Caradonna, op cit, p. 40.

148. Cited in Bruna Jordana and Paula Schuster, "Ambientalistas, pensando bem, somos nós," *Jornalismo Ambiental*, 27 Jun 2017.

149. Robin Wall Kimmerer, "Weaving..." op cit; Raymond Pierotti, *Indigenous Knowledge, Ecology, and Evolutionary Biology*, Routledge 2010.

150. Susan Smillie, "Tsunami, 10 Years On," *Guardian*, 10 Dec 2014.

151. Derek Elias, Soimart Rungmanee and Irwin Cruz, "The Knowledge that Saved the Sea Gypsies," *A World of Science*, vol. 3, no. 2, 2005.

Chapter 7: The Great Burning

1. *World Meterological Organization*, "WMO Provisional Statement on the Status of the Global Climate in 2016," 14 Nov 2016.
2. Omid Mazdiyasni et al., "Increasing Probability of Mortality During Indian Heat Waves," *Science Advances*, vol. 3, no. 6, 2017.
3. Caleb Scharf, "The Crazy Scale of Human Carbon Emission," *Scientific American*, 26 Apr 2017.
4. Ibid.
5. Ibid.
6. John Cook, "4 Hiroshima Bombs Worth of Heat Per Second," *Skeptical Science*, 1 Jul 2013.
7. Elizabeth Kolbert, "Can Carbon-Dioxide Removal Save the World?," *New Yorker*, 20 Nov 2017.
8. David Archer, *Global Warming: Understanding the Forecast*, John Wiley & Sons, 2012, p. 104.
9. For a brief guide to energy imbalance, see: James Hansen et al., "Earth's Energy Imbalance," *NASA—Science Briefs*, Jan 2012.
10. This is a common way of understanding heat balance. See: Harold H. Schobert, *Chemistry of Fossil Fuels and Biofuels*, Cambridge University Press, 2013, p. 453.
11. Christophe Bonneuil and Jean-Baptiste Fressoz, *The Shock of the Anthropocene*, Verso, 2016, p. 5.
12. For wider insights, see G. Thomas Farmer and John Cook, *Climate Change Science: A Modern Synthesis*, Springer, 2013, pp. 445–464.
13. John Cook et al., "Quantifying the consensus on anthropogenic global warming in the scientific literature," *Environmental Research Letters*, vol. 8, no. 2, 2013.
14. Rasmus E. Benestad et al., "Learning from Mistakes in Climate Research," *Theoretical and Applied Climatology*, vol. 126, nos. 3-4, 2016.
15. James Lawrence Powell, "The Consensus on Anthropogenic Global Warming Matters," *Bulletin of Science, Technology & Society*, vol. 36, no. 3, 2016.
16. See John Cook, "The Scientific Guide to Global Warming Skepticism," *Skeptical Science*, Dec 2010.
17. Ibid.
18. See Andreas Malm, *Fossil Capital*, Verso, 2016, p. 8; Joseph Romm, *Climate Change: What Everyone Needs to Know*, Oxford University Press, 2016, p. 8.
19. Zeke Hausfather, "Analysis: Why Scientists Think 100% of Global Warming is Due to Humans," *Carbon Brief*, 13 Dec 2017.
20. Zeke Hausfather and Robert McSweeney, "Q&A: How Do Climate Models Work?," *Carbon Brief*, 15 Jan 2017.

21. Hugues Goosse et al., *Introduction to Climate Dynamics and Climate Modelling*, Centre de recherche sur la Terre et le climat Georges Lemaître, 2010, pp. 59–85.
22. Zeke Hausfather and Robert McSweeney, op cit.
23. Ibid.
24. Ibid.
25. Mike Davis, "Has the Age of Chaos Begun?," *Tom Dispatch*, 6 Oct 2005.
26. Mario Giampetro, Kozo Mayumi and Alevgül H. Şorman, *Energy Analysis for a Sustainable Future*, Routledge, 2013, p. 311.
27. For a rich discussion of uncertainties in computer modeling, see: Arthur C. Petersen, *Simulating Nature*, CRC Press, 2012.
28. Joseph Romm, op cit, pp. 75–76, 136.
29. Leo Hickman, "The Carbon Brief Interview: Syukuro Manabe," *Carbon Brief*, 7 Jul 2015.
30. James C. McWilliams, "Irreducible Imprecision in Atmospheric and Oceanic Simulations," *Proceedings of the National Academy of Sciences*, vol. 104, no. 21, 2007.
31. Jeroen P. Van der Sluijs, "Uncertainty and Dissent in Climate Risk Assessment," *Nature and Culture*, vol. 7, no. 2, 2012.
32. Jeroen P. van der Sluijs, "Numbers Running Wild," *The Rightful Place of Science: Science on the Verge*, Arizona State University, 2016, pp. 166–167; Andrea Saltelli and Silvio Funtowicz, "When All Models Are Wrong," *Issues in Science and Technology*, vol. 30, no. 2, 2014.
33. Andrea Saltelli and Silvio Funtowicz, op cit.
34. Cited in: David Biello, "The Most Important Number in Climate Change," *Scientific American*, 30 Nov 2015.
35. Joseph Romm, op cit; Chris Mooney, "Scientists Keep Upping Their Projections for How Much the Oceans Will Rise This Century," *Washington Post*, 26 Apr 2017.
36. Josh Gabbatiss, "Global Sea Levels Could Rise 'Up to Five Metres' if Certain Antarctic Ice Sheets Melt," *Independent*, 14 Dec 2017.
37. Natalie M. Mahowald et al., "Are the Impacts of Land Use on Warming Underestimated in Climate Policy?," *Environmental Research Letters*, vol. 12, no. 9, 2017; Daniel L. Warner et al., "Carbon Dioxide and Methane Fluxes from Tree Stems, Coarse Woody Debris, and Soils in an Upland Temperate Forest," *Ecosystems*, vol. 20, no. 6, 2017; Julie Wolf, Ghassem R. Asrar and Tristram O. West, "Revised Methane Emissions Factors and Spatially Distributed Annual Carbon Fluxes for Global Livestock," *Carbon Balance and Management*, vol. 12, no. 1, 2017.
38. Yuan Suwen, Zhou Tailai and Li Rongde, "Northern China Chokes on Fake Emissions Data," *Caixin*, 6 Apr 2017.

39. Joseph Romm, op cit, p. 25.
40. Following Richard Alley, cited here: Juliet Eilperin, "Antarctic Ice Sheet Is Melting Rapidly," *Washington Post*, 3 Mar 2006.
41. Jeroen P. van der Sluijs, "Numbers Running Wild," op cit, pp. 159–160.
42. Patrick T. Brown and Ken Caldeira, "Greater Future Global Warming Inferred From Earth's Recent Energy Budget," *Nature*, vol. 552, no. 7683, 2017.
43. David Spratt and Ian Dunlop, "What Lies Beneath," *Breakthrough*, Sep 2017.
44. Ibid.
45. Cited in Joseph Romm, op cit, p. 140.

Chapter 8: Understanding Emissions:
Where, Who, What, When and How

1. Bobby Magill, "Natural Gas Emissions to Surpass Those of Coal in 2016," *Climate Central*, 30 Aug 2016.
2. Sophie Yeo, "Seven Charts Showing How Countries' Carbon Footprints Compare," *Carbon Brief*, 22 Aug 2016.
3. Joseph Romm, *Climate Change: What Everyone Needs to Know*, Oxford University Press, 2016, p. 23.
4. Ibid, p. 81.
5. H. Damon Matthews, "Quantifying Historical Carbon and Climate Debts Among Nations," *Nature Climate Change*, vol. 6, no. 1, 2016.
6. Andreas Malm, "Searching for the Origins of the Fossil Economy," *Verso Blogs*, 4 Dec 2005.
7. Andreas Malm, "Who Lit This Fire?," *Critical Historical Studies*, vol. 3, no. 2, 2016.
8. World Bank, "CO_2 Emissions, Metric Tons per Capita," nin.tl/WorldBank
9. Richard Heede, "Tracing Anthropogenic Carbon Dioxide and Methane Emissions to Fossil Fuel and Cement Producers, 1854–2010," *Climatic Change*, vol. 122, nos. 1-2, 2014.
10. Douglas Starr, "Just 90 Companies Are to Blame for Most Climate Change, This 'Carbon Accountant' Says," *Science Magazine*, 25 Aug 2016.
11. Richard Heede, op. cit.
12. Timothy Gore, "Extreme Carbon Inequality," Oxfam, 2015.
13. Ibid.
14. Andreas Malm, "Who Lit This Fire?," op. cit.
15. Anil Agarwal and Sunita Narain, *Global Warming in an Unequal World*, Centre for Science and Environment, 1991.

16. Andreas Malm and Alf Hornborg, "The Geology of Mankind?," *Anthropocene Review* vol. 1, no. 1, 2014.

17. Danny Dorling, "The Rich, Poor and the Earth," *New Internationalist*, 13 Jul 2017.

18. Personal interview with Asad Rehman.

19. Jeremy Leggett, *Carbon Wars*, Penguin, 2000, p. 3.

20. David Spratt and Ian Dunlop, op cit.

21. Chandrashekhar Dasgupta, "The Paris Agreement and Being Grateful for Small Mercies," *The Wire*, 16 Dec 2015.

22. Joseph Romm, op cit, p. 152.

23. James Hansen et al., "Assessing 'Dangerous Climate Change,'" *PloS one*, vol. 8, no. 12, 2013.

24. Klaus Bittermann et al., "Global Mean Sea-Level Rise in a World Agreed Upon in Paris," *Environmental Research Letters*, vol. 12, no. 12, 2017; IPCC, "Report on the Structured Expert Dialogue on the 2013–2015 Review," Bonn, Jun 2015.

25. Following Jeff Goodell, "The Doomsday Glacier," *Rolling Stone*, 9 May 2017.

26. Peter Hannam, "Warming Limit of 1.2 Degrees Needed to Save Great Barrier Reef," *The Age*, 2 Aug 2017; Chang-Eui Park et al., "Keeping Global Warming Within 1.5°C Constrains Emergence of Aridification," *Nature Climate Change*, vol. 8, 2018.

27. *The Emissions Gap Report 2017*, United Nations Environment Programme, 2017.

28. David Leonard Downie, Kate Brash and Catherine Vaughan, *Climate Change: A Reference Handbook*, ABC-CLIO, 2009, p. 12.

29. Peter J. Gleckler et al., "Industrial-era Global Ocean Heat Uptake Doubles in Recent Decades," *Nature Climate Change*, vol. 6, no. 4, 2016.

30. Joseph Romm, op cit, p. 6.

31. E. A. G. Schuur et al., "Climate Change and the Permafrost Carbon Feedback," *Nature*, vol. 520, no. 7546, 2015.

32. Joseph Romm, op cit, p. 81.

33. Trevor F. Keenan et al., "Recent Pause in the Growth Rate of Atmospheric CO_2 Due to Enhanced Terrestrial Carbon Uptake," *Nature Communications*, vol. 7, 2016.

34. Alex Whiting, "Climate Change May Turn Africa's Arid Sahel Green: Researchers," *Reuters*, 5 Jul 2017.

35. Oliver Geden and Andreas Löschel, "Define Limits for Temperature Overshoot Targets," *Nature Geoscience*, vol. 10, no. 12, 2017.

36. Jeremy L. Caradonna, *Sustainability: A History*, Oxford University Press, p. 243.

37. Frederick R. Steiner, *Human Ecology*, Island, 2016, p. 10.
38. Kate Marvel, "We Should Never Have Called It Earth," *On Being*, 1 Aug 2017.
39. Donella H. Meadows, *Thinking in Systems: A Primer*, Chelsea Green Publishing, 2008, p. 103.
40. Philippe Ciais et al., "Carbon and Other Biogeochemical Cycles," *Climate Change 2013: The Physical Science Basis*, Cambridge University Press, 2014, pp. 465–570.
41. Peter Erickson et al., "Assessing Carbon Lock-in," *Environmental Research Letters*, vol. 10, no. 8, 2015.
42. Steven J. Davis, Ken Caldeira and H. Damon Matthews, "Future CO_2 Emissions and Climate Change from Existing Energy Infrastructure," *Science*, vol. 329, no. 5997, 2010.
43. Sabrina Shankman, "Extreme Arctic Melt Is Raising Sea Level Rise Threat," *Inside Climate News*, 25 Apr 2017.
44. Joseph Romm, op cit, p. 29.
45. Thorsten Mauritsen and Robert Pincus, "Committed Warming Inferred from Observations," *Nature Climate Change*, vol. 7, no. 9, 2017.
46. We will ignite what climate scientist James Hansen described as a "a dynamic situation that is out of [human] control."
47. Joseph Romm, op cit, p. 148.
48. Ian Johnston, "World Has Three Years to Prevent Dangerous Climate Change, Warn Experts," *Independent*, 28 Jun 2017.
49. Brian Kahn, "Carbon Dioxide Is Rising at Record Rates," *Climate Central*, 10 Mar 2017.
50. Andrew P. Schurer et al., "Importance of the Pre-industrial Baseline for Likelihood of Exceeding Paris Goals," *Nature Climate Change*, vol. 7, no. 8, 2017.
51. David Spratt and Ian Dunlop, op cit.
52. Tobias Friedrich et al., "Nonlinear Climate Sensitivity and its Implications for Future Greenhouse Warming," *Science Advances*, vol. 2, no. 11, 2016.
53. Jiahua Pan, Jonathan Phillips and Ying Chen, "China's Balance of Emissions Embodied in Trade," *Oxford Review of Economic Policy*, vol. 24, no. 2, 2008.
54. Dale Jiajun Wen, "Climate Change and China," Focus on the Global South, 2009, p. 5.
55. Thomas O. Wiedmann et al., "The Material Footprint of Nations," *Proceedings of the National Academy of Sciences*, vol. 112, no. 20, 2015.
56. Karl W. Steininger et al., "Austria's Consumption-based Greenhouse Gas Emissions," *Global Environmental Change*, vol. 48, 2018.

57. Robert McSweeney and Rosamund Pearce, "Analysis: Just Four Years Left of the 1.5°C Carbon Budget," *Carbon Brief*, 5 Apr 2017.
58. Greg Muttitt, *The Sky's Limit*, Oil Change International, 2016.
59. Johan Rockström et al., "A Roadmap for Rapid Decarbonization," *Science*, Vol 355, no. 6331, 2017. An accessible discussion is available by Brad Plumer, "Scientists Made a Detailed 'Roadmap' for Meeting the Paris Climate Goals," *Vox*, 24 Mar 2017.
60. Kevin Anderson and Alice Bows, "Beyond 'Dangerous' Climate Change," *Philosophical Transactions of the Royal Society of London A: Mathematical, Physical and Engineering Sciences*, vol. 369, no. 1934, 2011.
61. Megan Darby, "Scientists: 1.5°C Warming Limit Means Fossil Fuel Phase-out by 2030," *Climate Home*, 7 Dec 2015.
62. Ed King, "New Fossil Fuel Plants Post-2017 Risk 2°C Warming Limit," *Climate Home*, 30 Mar 2016.
63. Adrian E. Raftery et al., "Less Than 2°C Warming by 2100 Unlikely," *Nature Climate Change* 7.9, 2017; Steven J. Davis, Ken Caldeira and H. Damon Matthews, "Future CO_2 Emissions and Climate Change from Existing Energy Infrastructure," *Science*, vol. 329, no. 5997, 2010. For an accessible discussion, see Glen Peters, "What is the Chance to Stay Below 2°C?," nin.tl/Peters
64. Glen Peters, "Can We Really Limit Global Warming to 'Well Below' Two Degrees Centigrade?," *Science Nordic*, 1 Oct 2017.
65. Kevin Anderson and Alice Bows, "A New Paradigm for Climate Change," *Nature Climate Change*, vol. 2, no. 9, 2012.
66. Alex Steffen, "The Last Decade and You," *The Nearly Now*, 6 Jun 2017.

Chapter 9: The Poverty of Wealth: Economics and Ecology

1. Arturo Escobar, "Una minga para el posdesarrollo," *Signo y pensamiento*, vol. 30, no. 58, 2011.
2. The work of Jason Hickel is useful in clarifying the politics around poverty-line statistics. See: Jason Hickel, *The Divide: A Brief Guide to Global Inequality and its Solutions*, William Heinemann, 2017.
3. Tharanga Yakupitiyage, "Progress on World Hunger Has Reversed," Inter Press Service, 3 Jul 2017.
4. Olga Khazan, "A Shocking Decline in American Life Expectancy," *The Atlantic*, 21 Dec 2017.
5. Ted Trainer, "Scrap the Conventional Model of Third World 'Development,'" *Resilience*, 5 Nov 2016.
6. Anis Chowdhury and Jomo Kwame Sundaram, "Growing Inequality under Global Capitalism," Inter Press Service, 4 May 2017.

7. Lorenzo Totaro and Giovanni Salzano, "Italy's Poor Almost Triple in a Decade Amid Economic Slumps," Bloomberg, 13 Jul 2017.
8. Facundo Alvaredo et al., "World Inequality Report 2018," World Inequality Lab, 2017.
9. Paul Buchheit, "Now Just Five Men Own Almost as Much Wealth as Half the World's Population," *Common Dreams*, 12 Jun 2017.
10. Isabel Ortiz and Matthew Cummins, "Global Inequality: Beyond the Bottom Billion," UNICEF, 2011.
11. Oxfam International, "Reward Work, Not Wealth," Oxfam Briefing Paper, Jan 2018.
12. Darrick Hamilton, "The Moral Burden on Economists," The Institute for New Economic Thinking, 13 Apr 2017.
13. Oxfam estimates that merely the amount of additional wealth accumulated by billionaires over a year period ($762bn) is enough to end extreme poverty seven times over. See: Oxfam International, "Reward Work, Not Wealth," Oxfam Briefing Paper, Jan 2018.
14. This comparison is widely invoked by Manfred Max-Neef.
15. For a discussion of their etymology see: Matthew Kurtz, "Economy and Ecology," *Open Learn*, 30 Jun 2006.
16. Cited in Mateo Aguado and Carlos Benítez Trinidad, "Entrevista a Alberto Acosta, economista y político," *Iberoamérica Social*, 17 Jun 2015.
17. Enrique Leff, "Decrecimiento o desconstrucción de la economía," *Polis: Revista Latinoamericana*, vol. 21, 2008.
18. Arnim Scheidel et al., "Ecological Distribution Conflicts as Forces for Sustainability," *Sustainability Science*, 2017.
19. Willi Haas, Simron Jit Singh and Annabella Musel, "Entropy," *EJOLT*, 2013.
20. Based on Herman Daly's rules, cited in Jeremy L. Caradonna, *Sustainability: A History*, Oxford University Press, 2014.
21. Angus Maddison, *The World Economy: Historical Statistics*, OECD, 2003.
22. Samuel Alexander, "Limits to Growth," *The Conversation*, 20 Apr 2016.
23. David Pilling, "Why It Is Time to Change the Way We Measure the Wealth of Nations," *Financial Times*, 8 Jan 2017.
24. David Bollier, "Re-imagining Value," *Heinrich Boell Foundation*, Mar 2017.
25. Jason Hickel, "Forget 'Developing' Poor Countries, It's Time to 'De-develop' Rich Countries," *Guardian*, 23 Sep 2015.
26. This relationship is known as the Easterlin Paradox.
27. Filipe Campante and David Yanagizawa-Drott, "Does Religion Affect

Economic Growth and Happiness," *Quarterly Journal of Economics*,
vol. 130, no. 2, 2015.

28. Jo Marchant, "Poorest Costa Ricans Live Longest," *Nature—News*,
3 Sep 2013.

29. Cited in Jason Hickel, *The Divide*, Heinemann, 2017, p. 141.

30. The work of K. William Kapp on cost-shifting is particularly interesting. See: Sebastian Berger, "K William Kapp's Theory of Social Costs
and Environmental Policy," *Ecological Economics*, vol. 67, no. 2, 2008.

31. Jason Moore and Raj Patel, *A History of the World in Seven Cheap
Things*, University of California Press.

32. See Ulrich Beck, *World at Risk*, Polity, 2009.

33. David Roberts, "None of the World's Top Industries Would Be Profitable If They Paid for the Natural Capital They Use," *Grist*, 17 Apr 2013.

34. World Bank, "Dominica Lost Almost All Its GDP Due to Climate
Change," World Bank News, 1 Dec 2015.

35. Soumya Sarkar, "India Is One of Most Vulnerable Nations to Climate
Change Impacts," *The Third Pole*, 12 Nov 2017.

36. James Cust, David Manley and Giorgia Cecchinato, "Unburnable
Wealth of Nations," *Finance & Development* 54.1, 2017.

37. Selina Williams and Bradley Olson, "Big Oil Companies Binge on
Debt," *Wall Street Journal*, 24 Aug 2016.

Chapter 10: The World at 1°C: A Guide to Climate Violence

1. Pedro Lemebel, "La inundación," 28 Apr 2006, nin.tl/Lemebel

2. Cited in Maddy Hodgson and Clare Hymer, "5 Reasons Climate
Change is a Feminist Issue," *Novara Media*, 11 Nov 2017.

3. Cited in Antonio Paz Cardona, "Hace 9 meses estudios advertían de
tragedia en Mocoa," *Semana Sostenible*, 1 Apr 2017.

4. Muhsin Al-Ramli, *The President's Gardens*, MacLehose Press, 2017.

5. Junot Díaz, "Apocalypse: What Disasters Reveal," *Boston Review*,
1 May 2011.

6. Cited in Stephen Jackson, "Un/natural Disasters, Here and There,"
Understanding Katrina: Perspectives from the Social Sciences, 2005.

7. Cited in Eric Holthaus, "Harvey and Irma Aren't Natural Disasters—
They're Climate Change Disasters," *Grist*, 11 Sep 2017.

8. Joseph Romm, *Climate Change: What Everyone Needs to Know*, Oxford
University Press, 2016, p. 51.

9. See: Virginia H. Dale et al., "Climate Change and Forest Distur-
bances," *BioScience*, vol. 51, no. 9, 2001; Nicola Jones, "Stark Evidence:
A Warmer World Is Sparking More and Bigger Wildfires," *Yale Environment 360*, 2 Oct 2017.

10. Ibid, pp. 49, 62.

11. Roz Pidcock, "In-depth: The Scientific Challenge of Extreme Weather Attribution," *Carbon Brief*, 11 Mar 2016.

12. Michael E. Mann et al., "Influence of Anthropogenic Climate Change on Planetary Wave Resonance and Extreme Weather Events," *Scientific Reports*, vol. 7, 2017.

13. Romm, op cit, p. 60.

14. Frank R. Rijsberman, "Every Last Drop," *Boston Review*, 1 Sep 2008.

15. Linus Unah, "Briefing: Nigerian Farmers Can't Fight Desertification Alone," *Irin News*, 14 Nov 2017.

16. Ken De Souva et al., "Vulnerability to Climate Change in Three Hot Spots in Africa and Asia," *Regional Environmental Change*, vol. 15, no. 5, Jun 2015; Richard S. J. Tol, "The Economic Impacts of Climate Change," *Review of Environmental Economics and Policy*, vol. 12, no. 1, 2018.

17. Chris Mooney, "Scientists Say the Pace of Sea Level Rise Has Nearly Tripled Since 1990," *Washington Post*, 22 May 2017; the 3.4 mm figure is based on the latest NASA figures.

18. The IPCC calls this sea level change.

19. William Colgan et al., "The Mind-bending Physics of Scandinavian Sea-level Change," *Science Nordic*, 3 Jan 2018.

20. Brian Kahn, "This Is How Climate Change Will Shift the World's Cities," *Climate Central*, 5 Jul 2017.

21. Daniel Macmillen Voskoboynik, "The Mocoa Massacre: Anatomy of a Tragedy Foretold," *The World at 1°C*, 4 Apr 2017.

22. Leslie Scism, "As Wildfires Raged, Insurers Sent in Private Firefighters to Protect Homes of the Wealthy," *Wall Street Journal*, 5 Nov 2017.

23. Mika Minio-Paluello, "Interviewing Mahienour: Revolution and Climate Change," *Platform*, 22 Jul 2014.

24. World Health Organization, "Gender, Climate Change and Health," 2014.

25. Brigitte Leoni, "UNISDR Head Calls for More Women in Disaster Risk Reduction," *UNISDR*, 8 Mar 2016.

26. Elain Enarson, "Women and Girls Last?: Averting the Second Post-Katrina Disaster," *Understanding Katrina: Perspectives from the Social Sciences*, 11 Jun 2006.

27. Margaret Alston, *Women and Climate Change in Bangladesh*. Routledge, 2015.

28. Chaman Pincha, *Indian Ocean Tsunami through The Gender Lens*, Oxfam America and NANBAN Trust, 2008.

29. Nathaniel E. Urama, Eric C. Eboh and Anthony Onyekuru, "Impact of Extreme Climate Events on Poverty in Nigeria," *Climate and Development*, 2017.

30. Amantha Perera, "As Water Vanishes in Sri Lanka, Baths—and Snakes—Present New Risks," *Reuters*, 15 Sep 2017.
31. Laurie Goering, "Silent Killer: Sweltering Planet Braces for Deadly Heat Shocks," Thomson Reuters Foundation, 19 Sep 2017.
32. CARE, "Rapid Gender Analysis: Cyclone Pam, Vanuatu," 7 April 2015, p. 8.
33. Sonia Narang, "Climate Change Drives Domestic Violence In Fiji," *Huffington Post*, 1 Jun 2017.
34. Justine Calma, "Climate Change Has Created a New Generation of Sex-trafficking Victims," *Quartz*, 2 May 2017; Gethin Chamberlain, "Why Climate Change Is Creating a New Generation of Child Brides," *Guardian*, 26 Nov 2017.
35. Manon Verchot, Indrani Basu and Joanna Plucinska, "Between the Dark Seas and Living Hell," *Huffington Post*, 1 Jul 2016.
36. Aravani is a term used in Tamil Nadu for *hijra*, "third-gender." Lori M. Hunter and Emmanuel David, "Displacement, Climate Change and Gender," *Migration and Climate Change*, 2011, pp. 306–330.
37. Chaman Pincha, op cit.
38. World Health Organization, "Gender, Climate Change and Health," 2014.
39. Ibid.
40. Gregor Wolbring, "A Culture of Neglect: Climate Discourse and Disabled People," *M/C Journal*, vol. 12, no. 4, 2009.
41. See: GPDD and World Bank, "The Impact of Climate Change on People with Disabilities," 8 Jul 2009, nin.tl/Disability
42. Airton Bostein, Valéria Vanda Azevedo de Lima and Angela Maria Abreu de Barros, "The Vulnerability of the Elderly in Disasters," *Ambiente & Sociedade*, vol. 17, no. 2, 2014.
43. Ibid.
44. C. Stöllberger, W. Lutz and Josef Finsterer, "Heat-related Side-effects of Neurological and Non-neurological Medication May Increase Heatwave Fatalities," *European Journal of Neurology*, vol. 16, no. 7, 2009.
45. UNICEF, "Climate Change and Children," Dec 2007, nin.tl/Children
46. Council on Environmental Health, "Global Climate Change and Children's Health," *Pediatrics*, vol. 136, no. 5, 2015.
47. Jason von Meding and Hang Thai TM, "In Vietnam Poverty and Poor Development, Not Just Floods, Kill the Most Marginalised," *The Conversation*, 29 Aug 2017.
48. Daniel Maxwell et al., "Facing Famine: Somali Experiences in the Famine of 2011," *Feinstein International Center*, October 2015, p. 14.

49. *Human Rights Watch*, "Hidden Apartheid: Caste Discrimination Against India's 'Untouchables,'" 2007.
50. Rina Chandran, "Heat and Drought Drive South India's Farmers from Fields to Cities," Thomson Reuters Foundation, 20 Sep 2017.
51. Tamara Steger (ed), *Making the Case for Environmental Justice in Central and Eastern Europe*, CEPL, HEAL & Coalition for Environmental Justice, 2007, p. 31.
52. Ethan D. Schoolman and Chunbo Ma, "Migration, Class and Environmental Inequality," *Ecological Economics*, vol. 75, 2012.
53. Aditi Roy Ghatak, "India's Floods Expose Poor Countries' Total Vulnerability to Climate Change," *Climate Home*, 1 Sep 2017.
54. Eric Holdeman, "Disaster Payouts—What the Numbers Show," *Emergency Management*, 7 Sep 2017.
55. Robert RM Verchick, *Facing Catastrophe: Environmental Action for a Post-Katrina World*, Harvard University Press, pp. 111–112.
56. This reading follows Kimberly Crenshaw's concept of intersectionality.
57. Naila Kabeer, *Reversed Realities*, Verso, 1994, p. 65.
58. Extracted from Daniel Macmillen Voskoboynik, "The Legal Limbo of Climate Refugees," *New Internationalist*, 14 Nov 2016.
59. Rob Nixon, *Slow Violence and the Environmentalism of the Poor*, Harvard University Press, 2011.
60. This is inspired in part by Reni Eddo-Lodge's framing of white privilege as an "absence of the consequences of racism"; Reni Eddo-Lodge, *Why I'm No Longer Talking to White People about Race*, Bloomsbury, 2018.

Chapter 11: A Plausible Future: Approaching Apocalypse

1. Cited in Charlotta Lomas, "Schellnhuber: 'Scientists Have to Take to the Streets' to Counter Climate Denial," *Deutsche Welle*, 15 Mar 2017.
2. Mike Davis, "The Necessary Eloquence of Protest," *The Nation*, 17 Mar 2009.
3. Mike Davis, "Living on the Ice Shelf," *Tom Dispatch*, 26 Jun 2008.
4. 'Antarctic Modeling Pushes Up Sea-Level Rise Projections," *Climate Central*, 13 Dec 2017.
5. Robert M. DeConto and David Pollard, "Contribution of Antarctica to Past and Future Sea-level Rise," *Nature*, vol. 531, no. 7596, 2016.
6. Margaret A. Palmer et al., "Climate Change and the World's River Basins," *Frontiers in Ecology and the Environment*, vol. 6, no. 2, 2008.
7. James Jeffrey, "Falling Between the Sun-Scorched Gaps," *Inter Press Service*, 10 May 2017.

8. Yangyang Xu and Veerabhadran Ramanathan, "Well Below 2°C," *Proceedings of the National Academy of Sciences*, 2017.

9. Anthony D. Barnosky et al., "Approaching a State Shift in Earth's Biosphere," *Nature*, 486, vol. 7401, 2012.

10. Richard Mahapatra, "Why I Mourn Dussehra: Climate Change and Death of Seasons," *Down to Earth*, 29 Sep 2017.

11. Gleb Raygorodetsky, "They Migrate 800 Miles a Year," *National Geographic*, Oct 2017.

12. Ning Lin and Kerry Emanuel, "Grey Swan Tropical Cyclones," *Nature Climate Change*, vol. 6, no. 1, 2016.

13. Vernon Loeb, "Harvey Should be the Turning Point in Fighting Climate Change," *Washington Post*, 29 Aug 2017.

14. Angela Fritz, "This City in Alaska Is Warming So Fast, Algorithms Removed the Data Because It Seemed Unreal," *Washington Post*, 12 Dec 2017.

15. Stan Cox and Paul Cox, *How the World Breaks*, The New Press, 2016, p. 37; Eric Holthaus, "Let It Go: The Arctic Will Never Be Frozen Again," *Grist*, 18 Dec 2017.

16. Jonathan N. Pauli et al., "The Subnivium: A Deteriorating Seasonal Refugium," *Frontiers in Ecology and the Environment*, vol. 11, no. 5, 2013.

17. Medical journal *The Lancet* described climate change as "the number-one threat to global public health in the 21st century."

18. Michaeleen Doucleff, "Anthrax Outbreak in Russia Thought To Be Result of Thawing Permafrost," *NPR*, 3 Aug 2016.

19. Christian Bogdal et al., "Blast from the Past," *Environmental Science & Technology*, vol. 43, no. 21, 2009; Chris Mooney, "The Arctic Is Full of Toxic Mercury, and Climate Change Is Going to Release It," *Washington Post*, 5 Feb 2018.

20. Filka Sekulova and J. C. J. M. Van den Bergh, "Floods and Happiness," *Ecological Economics*, vol. 126, 2016.

21. Cited in Matt Smith, "Impact of Climate Change on Public Health Gets Renewed Focus," *Seeker*, 12 Feb 2017.

22. John Fischer, "Climate Change Isn't Just Impacting Crops; It's Taking a Physical and Psychological Toll on Farmers," *Medical Daily*, 10 Dec 2014.

23. P. Sainath, "Maharashtra Crosses 60,000 Farm Suicides," 15 Jul 2014, nin.tl/Sainath.

24. Jitendra Choubey, "Every Hour, a Farmer Committed Suicide in 2015," *Down to Earth*, 2 Jan 2017.

25. Surendra P. Gangan, "2,414 Farmers Commit Suicide in Maharashtra

from Jan to Oct; No Dip Despite Loan Waivers," *Hindustan Times*, 17 Nov 2015.

26. Jyoti Shelar, "The Silent Sufferers: On Maharashtra Farmer Suicides," *The Hindu*, 17 Feb 2018.

27. Cited in: Keith Schneider, "Worst Drought in 140 Years Leads to Farmer Deaths, Riots, Policy Impasse in Cauvery Delta," *New Security Beat*, 17 Apr 2017.

28. Livia Albeck-Ripka, "Why Lost Ice Means Lost Hope for an Inuit Village," *New York Times*, 25 Nov 2017.

29. Oliver Milman, "A Third of the World Now Faces Deadly Heatwaves as Result of Climate Change," *Guardian*, 19 Jun 2017.

30. Gayathri Vaidyanathan, "Killer Heat Grows Hotter Around the World," *Scientific American*, 6 Aug 2015.

31. Camilo Mora, et al., "Twenty-Seven Ways a Heat Wave Can Kill You: Deadly Heat in the Era of Climate Change, *Circulation: Cardiovascular Quality and Outcomes*, vol. 10, no. 11, 2017.

32. Bob Berwyn, "Heat Waves Creeping Toward a Deadly Heat-Humidity Threshold," *Inside Climate News*, 3 Aug 2017.

33. Jeremy S. Pal and Elfatih AB Eltahir, "Future Temperature in Southwest Asia Projected to Exceed a Threshold for Human Adaptability," *Nature Climate Change*, vol. 6, 2016.

34. Fred Pearce, "Global Warming Could Make Hajj Impossible Later This Century," *New Scientist*, 26 Oct 2015.

35. Oliver Milman, op cit.

36. Joseph Romm, *Climate Change: What Everyone Needs to Know*, Oxford University Press, 2016, p. 35.

37. Stan Cox and Paul Cox, op cit, p. 33, 38.

38. Cited in Manipadma Jena, "Killer Heatwaves Set for Dramatic Rise, Researchers Warn," *Reuters*, 19 Jun 2017.

39. Rina Chandran, "Heat and Drought Drive South India's Farmers From Fields to Cities," *Reuters*, 20 Sep 2017.

40. Ibid.

41. Nick Meynen, "The New Oil?," *Scroll*, 4 May 2017.

42. Peter Schwartzstein, "What Will Happen If the World No Longer Has Water?," *Newsweek*, 22 Nov 2017.

43. Dale Wen, "Climate Change, Energy and China," *Sparking a Worldwide Energy Revolution*, AK Press 2010, p. 134.

44. Mike Davis, "The Coming Desert," *New Left Review*, vol. 97, Jan-Feb 2016.

45. Joseph Romm, op cit, p. 99.

46. Ibid, p. 127.

47. United Nations, "Water and Disaster Risk," nin.tl/Water

48. Sophie Hares, "Early Warning Systems Still Missing in 100 Countries, UN says," Reuters, 24 May 2017.
49. Ben Lih Yi, "Asia Must Invest More in Disaster Risk Reduction—Red Cross," Thomson Reuters Foundation, 10 Aug 2016.
50. Joseph Romm, op cit, p. 137.
51. Lauren Zanolli, "Louisiana's Vanishing Island," *Guardian*, 15 Mar 2016.
52. Liz Koslov, "The Case for Retreat," *Public Culture*, vol. 28, no. 2, 2016.
53. Baher Kamal, "Climate Victims—Every Second, One Person Is Displaced by Disaster," *Inter Press Service*, 27 Jul 2016.
54. Edward Wong, "Resettling China's 'Ecological Migrants,'" *New York Times*, 25 Oct 2016.
55. Nellie Peyton, "Senegal City Races to Move Families as Sea Swallows Homes," *Place*, 3 Apr 2018; Richard Stone, "Cuba Embarks on a 100-Year Plan to Protect Itself from Climate Change," *Science*, 10 Jan 2017.
56. Linda Pressly, "The Island People with a Climate Change Escape Plan," *BBC news*, 21 Sep 2017.
57. Natalie Suckall, Evan Fraser and Piers Forster, "Reduced Migration Under Climate Change," *Climate and Development*, vol. 9, no. 4, 2017.
58. On reluctance, see: Ma Laurice Jamero et al., "Small-island Communities in the Philippines Prefer Local Measures to Relocation in Response to Sea-level Rise," *Nature Climate Change*, vol. 7, no. 8, 2017.
59. See: Josh Toussaint-Strauss et al., "Some Don't Have Bodies to Bury," *Guardian*, 14 Dec 2017.
60. Daisy Dunne, "Climate Change 'Could Double' the Number of Droughts in Jordan by 2100," *Carbon Brief*, 30 Aug 2017.
61. Elisabeth Vallet, "Border Walls Are Ineffective, Costly and Fatal—But We Keep Building Them," *The Conversation*, 4 Jul 2017.
62. Robtel Neajai Pailey, "Legal Invisibility Was the Best Thing to Happen to Me," *African Arguments*, 31 May 2017.
63. Uma Kothari, "Political Discourses of Climate Change and Migration," *Geographical Journal*, vol. 180, no. 2, 2014.
64. Edward Wong, op cit.
65. Danilo Urrea, "La sed del carbón. Causas estructurales de la sequía en La Guajira," *CENSAT Agua Viva*, nin.tl/Guajira.
66. Stan Cox and Paul Cox, op cit, p. 89; Naomi Klein, *The Shock Doctrine: The Rise of Disaster Capitalism*, Macmillan, 2007; John C. Mutter, *The Disaster Profiteers*, St. Martin's, 2015.
67. Lauren H. Derby, *The Dictator's Seduction*, Duke University Press 2009, pp. 66–90.
68. See: Timothy Snyder, *Black Earth*, Bodley Head, 2015.

69. Andy Horowitz, "The Racial Strife That Can Blow in With a Hurricane," *Washington Post*, 25 Aug 2017.
70. Romain Felli, "Climate Politics in the Long Run," *Entitle Blog*, 25 July 2017.
71. Joachim Radkau, *Nature and Power*, Cambridge University Press, 2008, pp. 246–247.
72. Lisa Sun-Hee Park and David N. Pellow, *The Slums of Aspen*, NYU Press, 2011; Joyce E. Chaplin, "Is Greatness Finite?," *Aeon*, 26 Jan 2017.
73. Richard Martyn-Hemphill and Henrik Pryser Libellmarch, "Who Owns the Vikings?," *New York Times*, 17 Mar 2018.
74. Solomon M. Hsiang and Marshall Burke, "Climate, Conflict, and Social Stability: What Does the Evidence Say?," *Climatic Change*, vol. 123, no. 1, 2014.
75. Peter Schwartzstein, "Climate Change and Water Woes Drove ISIS Recruiting in Iraq," *National Geographic*, 14 Nov 2017.
76. Sebastian van Baalen and Malin Mobjörk, "Climate Change and Violent Conflict in East Africa," *International Studies Review*, vix 043, 10 Nov 2017.
77. Wario R. Adano et al., "Climate Change, Violent Conflict and Local Institutions in Kenya's Drylands," *Journal of Peace Research*, vol. 49, no. 1, 2012; L. Jen Shaffer, "An Anthropological Perspective on the Climate Change and Violence Relationship," *Current Climate Change Reports*, vol. 3, no. 4, 2017.
78. Eric Holt Gimenez, "Climate Violence and the Criminalization of Hunger," *Huffington Post*, 7 Apr 2017.
79. Sebastian van Baalen and Malin Mobjörk, op cit.
80. Jitendra Choubey "Africa: How Climate Change Turned Two of the World's Oldest Tribes into Enemies," *Down to Earth*, 31 Oct 2017.
81. See Jan Selby and Clemens Hoffman (eds), *Rethinking Climate Change, Conflict and Security*, Routledge, 2016.
82. Fram Dinshaw, "Climate Change Opened 'Gates of Hell,' in Syria: Al Gore," *National Observer*, 9 Jul 2015.
83. For more context, see: Daniel Macmillen Voskoboynik, "No, Climate Change Did Not 'Cause' the Syrian War," *Pulse*, 16 Dec 2016.
84. Francesca de Châtel, "The Role of Drought and Climate Change in the Syrian Uprising," *Middle Eastern Studies*, vol. 50, no. 4, 2014.
85. "Syrian Writers, Artists, and Journalists Speak Out Against US and Russian Policy," *The Nation*, 21 Sep 2016.
86. Joseph Romm, op cit, pp. 137–138.
87. Lisa Friedman and Gayathri Vaidyanathan, "Irreversible Climate Change Would Result from Continued Inaction," *Scientific American*, 28 Oct 2014.

88. Rachel Warren, "The Role of Interactions in a World Implementing Adaptation and Mitigation Solutions to Climate Change," *Philosophical Transactions of the Royal Society of London: Mathematical, Physical and Engineering Sciences*, vol. 396, no. 1934, 2011.

89. Kurt M. Campbell et al., *The Age of Consequences*, Centre for Strategic and International Studies, 2007.

90. Cited in Todd Miller, *Storming the Wall*, City Lights Books, 2017.

91. CNA Corporation, "National Security and the Threat of Climate Change," 2007, nin.tl/NationalSecurity.

92. James E. Hansen and Makiko Sato, "Earth's Climate History: Implications for Tomorrow," *NASA-Briefs*, Jul 2011.

93. Peter U. Clark et al., "Consequences of 21st-century Policy for Multi-millennial Climate and Sea-level Change," *Nature Climate Change*, vol. 6, no. 4, 2016.

Chapter 12: A Possible Future: The World We Can Win

1. Boaventura de Sousa Santos, *De las dualidades a las ecologías*, REMTE, 2012, p. 7.

2. Cited in Giorgos Kallis, *In Defense of Degrowth: Opinions and Minifestos*, 2017, p. 62.

3. Cited in Lisa VeneKlasen, "¡Berta Vive! Lecciones de la resistencia en Honduras," *Open Democracy*, 8 Mar 2017.

4. Imre Szeman and Petrocultures Research Group, *After Oil*, West Virginia University Press, 2016.

5. As chemist Ilya Prigogine and philosopher Isabelle Stengers note, "has a pluralistic, complex character." See: Ilya Prigonine and Isabelle Stengers, *Order out of Chaos*, Verso, 2018.

6. Although widely ascribed to Unger, this is actually a paraphrasing by Lee Smolin, *The Life of the Cosmos*, Oxford University Press, 1997, p. 258.

7. Wade Davis, *The Wayfinders*, House of Anansi Press, 2009.

8. Federico Demaria and Ashish Kothari, "The Post-Development Dictionary Agenda: Paths to the Pluriverse," *Third World Quarterly*, vol. 38, no. 12, 2017; Giacomo D'Alisa, Federico Demaria and Giorgos Kallis (eds), *Degrowth: A Vocabulary for a New Era*, Routledge, 2014; Munyaradzi Felix Murove, "The Shona Ethic of Ukama with Reference to the Immortality of Values," *Mankind Quarterly*, vol. 48, no. 2, 2007.

9. Irene León, *Sumak Kawsay: buen vivir y cambios civilizatorios*, Fedaeps, 2010; Julieta Paredes, *Hilando fino*, Comunidad Mujeres Creando Comunidad, 2010.

10. Nancy Cartwright, "Presidential Address: Will This Policy Work for You?" *Philosophy of Science*, vol. 79, no. 5, 2012.

11. James Milner et al., "Home Energy Efficiency and Radon Related Risk of Lung Cancer," *Bmj*, vol. 348, 2014.

12. Maria Kaika, "Our Sustainability Is Someone Else's Disaster," *Green European Journal*, 5 Jan 2018.

13. This phenomenon is known as green gentrification.

14. Li Jing, "No Heating at −6C," *Climate Home*, 4 Dec 2017; Claudia Ciobanu, "Dutch Unions Back Coal Phaseout but Demand Compensations," *Just Transition*, 20 Feb 2018.

15. Jennifer R. Marlon et al., "Long-term Perspective on Wildfires in the Western USA," *Proceedings of the National Academy of Sciences*, vol. 109, no. 9, 2012.

16. Stan Cox, "Is Greenhouse Warming a Good Pretext for Selling Driverless Cars?," *Resilience*, 6 Nov 2017.

17. Sara Ahmed, *The Promise of Happiness*, Duke University Press, 2010; Barbara Ehrenreich, *Bright-Sided*, Metropolitan Books, 2009; Stan Cox and Paul Cox, *How the World Breaks*, New Press, 2016.

18. Linus Karlsson et al., "'Triple Wins' or 'Triple Faults'?," *Journal of Peasant Studies*, vol. 45, no. 1, 2018.

19. Robert W. Kates, William R. Travis and Thomas J. Wilbanks, "Transformational Adaptation When Incremental Adaptations to Climate Change Are Insufficient," *Proceedings of the National Academy of Sciences*, vol. 109, no. 19, 2012.

20. Chad Gillis, "Seawalls Across Southwest Florida Crumbling in Wake of Irma," *News Press*, 18 Sep 2017.

21. Manoj Roy, David Hulme and Joseph Hanlon, "It's Too Early to Talk About Climate Change Refugees in Bangladesh," *The Conversation*, 11 Nov 2016. For wider discussion, see: Alexandre Magnan, "Avoiding Maladaptation to Climate Change," *Surveys and Perspectives Integrating Environment and Society*, vol. 7, no. 1, 2014.

22. Patrick McCully, "Before the Deluge," *IRN Dams, Rivers and People Report*, 2007.

23. Jeff Goodell, "Rising Waters," *Yale Environment 360*, 5 Dec 2017

24. Ibid.

25. Wendell Berry, *The Way of Ignorance And Other Essays*, Counterpoint, 2006, p. 21.

26. Mario Giampietro and Kozo Mayumi, *The Biofuel Delusion*, Routledge, 2009.

27. Arthur Neslen, "New Airplane Biofuels Plan Would 'Destroy Rainforests,' Warn Campaigners," *Guardian*, 12 Oct 2017.

28. Barbara Unmüssig, "The Geoengineering Fallacy," Project Syndicate, 12 Oct 2017.

29. Abby Rabinowitz and Amanda Simson, "The Dirty Secret of the World's Plan to Avert Climate Disaster," *Wired*, 12 Oct 2017.
30. Action Aid, "Caught in the Net," Jun 2015.
31. Niall MacDowell et al., "The Role of CO_2 Capture and Utilization in Mitigating Climate Change," *Nature Climate Change*, vol. 7, no. 4, 2017.
32. Global CCS Institute, "Projects Database," nin.tl/CCS
33. Andy Skuce, "The Quest for CCS," *Corporate Knights*, Winter 2016.
34. Ibid.
35. Tim Radford, "World's Young Face $535 Trillion Bill for Climate," *Climate News Network*, 19 Jul 2017.
36. Biofuelwatch, Heinrich Böll Foundation and ETC Group, "The Big Bad Fix," 2017.
37. Raymond T. Pierrehumbert, "The Trouble With Geoengineers 'Hacking the Planet,'" *Bulletin of the Atomic Scientists*, 23 Jun 2017.
38. Biofuelwatch, Heinrich Böll Foundation and ETC Group, op cit.
39. Ibid.
40. Tina Lasisi and Mark D. Shriver, "Focus on African Diversity Confirms Complexity of Skin Pigmentation Genetics," *Genome Biology*, vol. 19, no. 1, 2018.
41. Andrew Zolli, "Toward a Contemplative Ecology," *On Being*, 25 Apr 2017.
42. Sunita Narain, *First Food: A Taste of India's Biodiversity*, CSE, 2013.
43. This is known as the principle of subsidiarity.
44. Nnimmo Bassey, "COP21 Agreed to a Climate-changed World," *New Internationalist*, 15 December 2015.
45. Cited in Frank McDonald, "Climate Deal Reaction," *Irish Times*, 12 Dec 2015.
46. The work of Cornelius Castoriadis is path-breaking when considering ecology and democracy. Also see: Mauro Bonaiuti, "Growth and Democracy," *Futures*, vol. 44, no. 6, 2012; Jean-Louis Martin, Virginie Maris, and Daniel S. Simberloff, "The Need to Respect Nature and its Limits Challenges Society and Conservation Science," *Proceedings of the National Academy of Sciences*, vol. 113, no. 22, 2016.
47. Filka Sekulova, Giorgos Kallis and François Schneider, "Climate Change, Happiness and Income from a Degrowth Perspective," *Handbook on Growth and Sustainability*, Edward Elgar Publishing, 2017.
48. This idea is rooted in the *Leap Manifesto*, "A Call for Canada Based on Caring for the Earth and One Another," Sep 2015.
49. Erik Leipoldt, "People with Disabilities in a Disabled World," *Physical Disability Council of Australia Annual Forum*, Nov 2004. Accessible at: nin.tl/Leipoldt

50. Neera Shrestha Pradhan, "Reaching the Most Vulnerable Across the Border," *International Centre for Integrated Mountain Development*, 12 Aug 2017.

51. Lisa Buggy and Karen Elizabeth McNamara, "The Need to Reinterpret 'Community' for Climate Change Adaptation," *Climate and Development*, vol. 8, no. 3, 2016; Katherine Browne, "Who Lives Here?," *Natural Hazards Center*, 23 Oct 2017; Isabella Troconis, "Gas Butano vs. Energía Solar," nin.tl/Butano

52. Nick Miroff, "A Flood of Problems," *Washington Post*, 7 Aug 2017.

53. Cited in Donella H. Meadows, *Thinking in Systems: A Primer*, Chelsea Green Publishing, 2008, p. 103.

54. Deborah Coen, "Climate Change and the Humanities," *Items*, 9 Jan 2018.

55. As Alex Steffen writes, "Winning Slowly Is Basically the Same Thing as Losing Outright." See Steffen's writing on predatory delay at: Alex Steffen, "The Last Decade and You," *The Nearly Now*, 6 Jun 2017.

56. Brian Clark Howard, "India Plants 50 Million Trees in One Day, Smashing World Record," *National Geographic*, 18 Jul 2016.

57. *Mike Davis, "Home-Front Ecology,"* Sierra Magazine, Jul-Aug 2007; Christopher Grainger, "Why We Need a 'Space Race' Approach to Saving the Planet," *The Conversation*, 2 Dec 2015; Giorgos Kallis, "In Defence of Degrowth," *Ecological Economics*, vol. 70, no. 5, 2011.

58. Luke Sussams and James Leaton, "Expect the Unexpected," *Carbon Tracker & Grantham Institute of Imperial College*, 2017, p. 23.

Chapter 13: A Mosaic of Alternatives

1. Bayo Akomolafe, "Out of the Box," nin.tl/BayoAkomolafe

2. Tim Jackson, *Prosperity Without Growth*, Earthscan 2009; Peter Victor, *Managing Without Growth*, Edward Elgar, 2008.

3. Ana Esther Ceceña, "Pensar la vida y el futuro de otra manera," *Sumak Kawsay/Buen Vivir y cambios civilizatorios*, FEDAEPS, 2010, p. 84.

4. Deborah S. Rogers et al., "A Vision for Human Well-being," *Current Opinion in Environmental Sustainability*, vol. 4, no. 1, 2012.

5. Manfred Max-Neef, "The World on a Collision Course and the Need for a New Economy," *AMBIO: A Journal of the Human Environment*, vol. 39, no. 3, 2010.

6. Isabel Carvalho et al., *Línea de Dignidad*, Programa Cono Sur Sustentable, 2002.

7. Manfred Max-Neef, op cit.

8. University of Leeds, "A Good Life For All Within Planetary Boundaries," nin.tl/Boundaries

9. Kate Raworth, *Doughnut Economics*, Random House Business Books, 2017.
10. Anna Bosch, Cristina Carrasco and Elena Grau, "Verde que te quiero violeta," *La Historia Cuenta*, Ediciones El Viejo Topo, 2005.
11. See Kevin Anderson, cited in Kate Aronoff, "No Third Way for the Planet," *Jacobin*, 5 Oct 2017.
12. Marjorie Kelly, *Owning Our Future*, Berrett-Koehler Publishers, 2012.
13. Michael Buehler, Edmundo Werna and Mark Brown, "More Than 2 Million People Die at Work Each Year," *World Economic Forum*, 23 Mar 2017.
14. Mayumi Hayashi, "Japan's Fureai Kippu Time-banking in Elderly Care," *International Journal of Community Currency Research*, vol. 16, 2012.
15. Enrique Leff, "Ecotechnological Productivity," *Information (International Social Science Council)*, vol. 25, no. 3, 1986.
16. Rachel Banning-Lover, "How to Stop Deforestation," *Guardian*, 4 Apr 2017.
17. Adam Morton, "Miners Receive Twice as Much in Tax Credits as Australia Spends on Environment," *Guardian*, 1 Feb 2018.
18. Thomas Marois, "How Public Banks Can Help Finance a Green and Just Energy Transformation," *Transnational Institute*, Nov 2017.
19. Eduardo Gudynas, "Sentidos, opciones y ámbitos de las transiciones al postextractivismo," *Más allá del desarrollo*, 2011; Vicente Sotelo and Pedro Francke, "¿Es económicamente viable una economía post extractivista en el Perú," *Transiciones*, Ediciones del CEPES, 2011, pp. 115–142.
20. Hazel Healy, "Bad Education: Why Our Systems Need Fixing," *New Internationalist*, Sep 2017.
21. Kevin Anderson and Alice Bows, "Beyond 'Dangerous' Climate Change," *Philosophical Transactions of the Royal Society of London A: Mathematical, Physical and Engineering Sciences* 369.1934, 2011.
22. Tim Kasser, "The Good Life or the Goods Life?," *Positive Psychology in Practice*, Wiley, 2004.
23. Lorenzo Fioramonti, *Wellbeing Economy*, Macmillan, 2017.
24. Daniel W. O'Neill et al., "A Good Life for All Within Planetary Boundaries," *Nature Sustainability*, vol. 1, no. 2, 2018.
25. Leah Burrows, "Smoke from 2015 Indonesian Fires May Have Caused 100,000 Premature Deaths," *Harvard John A. Paulson School of Engineering and Applied Sciences*, 19 Sep 2016.
26. Mongabay Haze Beat, "15 Fire-linked Firms Escape Prosecution in Indonesia's Riau," *Mongabay*, 28 Jul 2016.

27. Jennifer Chu, "Study: Volkswagen's Excess Emissions Will Lead to 1,200 Premature Deaths in Europe," MIT News, 3 March 2017.
28. Nick Visser, "Exxon Mobil Fined $21 Million for Violating Clean Air Act 16,386 Times," Huffington Post, 27 Apr 2017.
29. Rachel B. Smith, et al., "Impact of London's Road Traffic Air and Noise Pollution on Birth Weight," BMJ, vol. 359, 2017.
30. Annie Leonard, The Story of Stuff, Constable, 2016, p. 104.
31. Cited in Derrick Jensen, The Culture of Make Believe, Chelsea Green Publishing, 2004, p. 285.
32. Fred H. Besthorn, "Restorative Justice and Environmental Restoration," Journal of Societal and Social Policy, vol. 3, no. 2, 2004.
33. Chris Arsenault, "Why Are Most of the World's Hungry People Farmers?," Thomson Reuters Foundation, 27 May 2015.
34. Oliver Milman, "Earth Has Lost a Third of Arable Land in Past 40 Years, Scientists Say," Guardian, 2 Dec 2015.
35. Joachim Radkau, Nature and Power, Cambridge University Press, 2008, p. 251.
36. Lauren C. Ponisio et al., "Diversification Practices Reduce Organic to Conventional Yield Gap," Proc R Soc B, vol. 282, no. 1799, 2015; John P. Reganold and Jonathan M. Wachter, "Organic Agriculture in the 21st Century," Nature Plants, vol. 2, no. 2, 2016.
37. Jules N. Pretty et al., "Resource-conserving Agriculture Increases Yields in Developing Countries," Environmental Science & Technology, vol. 40, no. 4, 2006.
38. Pat Mooney and Nnimmo Bassey, "The Road to Food Sovereignty," New Internationalist, 14 Dec 2017.
39. Charles Eisenstein, "Opposition To GMOs Is Neither Unscientific Nor Immoral," Huffington Post, 8 Jan 2018.
40. Kathryn Rawe, "A Life Free From Hunger," Save the Children, 2012.
41. Lovish Garg, "Beyond the Green Fields of Punjab Lies a Mounting Agrarian Distress," The Wire, 28 Aug 2017.
42. Paméla Rougerie, "Quiet Epidemic of Suicide Claims France's Farmers," New York Times, 20 Aug 2017; Debbie Weingarten, "Why Are America's Farmers Killing Themselves in Record Numbers?," Guardian, 6 Dec 2017; Baba Umar, "India's Shocking Farmer Suicide Epidemic," Al Jazeera, 18 May 2015.
43. Philip Lymbery, Farmageddon, Bloomsbury, 2014.
44. Patricia Aguirre, "Alternativas a la crisis global de la alimentación," NUSO, Mar-Apr 2016.
45. Saraswathy Nagarajan, "Indigenous Rice Varieties Make a Comeback," The Hindu, 11 Jan 2018.
46. David R. Montgomery, Dirt: The Erosion of Civilizations, University of California Press, 2012, pp. xiv and 23.

47. Paul Hawken, *Drawdown: The Most Comprehensive Plan Ever Proposed to Roll Back Global Warming*, Penguin, 2017.
48. Braulio Machín-Sosa et al., *Revolución Agroecológicam*, ANAP, 2010; Eric Holt-Giménez, "Measuring Farmers' Agroecological Resistance after Hurricane Mitch in Nicaragua," *Agriculture, Ecosystems and Environment*, vol. 93, 2002.
49. Jonathan Sanderman, Tomislav Hengl and Gregory J. Fiske, "Soil Carbon Debt of 12,000 Years of Human Land Use," *Proceedings of the National Academy of Sciences*, vol. 114, no. 36, 2017.
50. Andreas Gattinger et al., "Enhanced Top Soil Carbon Stocks Under Organic Farming," *Proceedings of the National Academy of Sciences*, vol. 109, no. 44, 2012; Rattan Lal, "Soil Carbon Sequestration Impacts on Global Climate Change and Food Security," *Science*, vol. 304, no. 5677, 2004.
51. Baher Kamal, "Agony of Mother Earth (II) World's Forests Depleted for Fuel," *Inter Press Service*, 19 May 2017.
52. Marco Springmann et al., "Analysis and Valuation of the Health and Climate Change Cobenefits of Dietary Change," *Proceedings of the National Academy of Sciences*, vol. 113, no. 15, 2016.
53. Joseph Romm, *Climate Change: What Everyone Needs to Know*, Oxford University Press, 2016, p. 247.
54. Jenny Gustavsson et al., "Global Food Losses and Food Waste," FAO, 2011.
55. FAO, "Food Wastage Footprint and Climate Change," nin.tl/FoodWaste
56. Ashish Kothari, "Radical Ecological Democracy," *The Great Transition*, Jul 2014.
57. Anil Bhattarai, "Transcending the Focus on Agrarian Sector," *Food Sovereignty: A Critical Dialogue*, Sep 2013, nin.tl/Anil
58. David R. Montgomery, op cit, p. 23.
59. Darío Aranda, "La contaminación en las aulas," *Página 12*, 15 Apr 2014; Pablo Plotkin, "Retratos de la era tóxica," *Rolling Stone*, 29 Sep 2016.
60. Meena Menon, "A Litany of Miseries," *Scroll*, 12 Nov 2017.
61. Ryan Rafai, "UN: 200,000 Die Each Year from Pesticide Poisoning," *Al Jazeera*, 8 Mar 2017.
62. Damian Carrington, "Sixth Mass Extinction of Wildlife Also Threatens Global Food Supplies," *Guardian*, 26 Sep 2017.
63. Colin K. Khoury et al., "Increasing Homogeneity in Global Food Supplies and the Implications for Food Security," *Proceedings of the National Academy of Sciences*, vol. 111, no. 11, 2014.
64. Mario Giampetro, "Agricultura i alimentació per a un futur sostenible," *Defensem Catalunya del Canvi Climàtic*, 3 Mar 2018, nin.tl/Giampetro

65. Víctor M. Toledo and Narciso Barrera-Bassols, *La memoria biocultural*, Icaria Editorial, 2008.
66. José Luis Vivero Pol, "Food as Commons or Commodity?," *Sustainability*, vol. 9, no. 3, 2017.
67. Sylvia Kay, "Land Grabbing and Land Concentration in Europe," Transnational Institute, Dec 2016, p. 14.
68. Arantxa Guereña, "What the Latest Agricultural Census Reveals About Land Distribution in Colombia," Oxfam International, 10 Jul 2017.
69. Richard Mahapatra, "Only 15% Landholders Earn 91% of Total National Income," *Down to Earth*, 16 Jan 2018.
70. Michel Bauwens, "The History and Evolution of the Commons," *Commons Transition*, 19 Sep 2017.
71. David Bollier, "Commoning as a Transformative Social Paradigm," *The Next System Project*, 28 Apr 2016.
72. Joydeep Gupta, "South Asia Braces for Intense Heat This Summer," *The Third Pole*, 5 Apr 2017; Maria E. Fernandez-Gimenez et al., "Lessons from the Dzud," *World Development*, vol. 68, 2015.
73. This idea is explored in Stephanie LeMenager, *Living Oil*, Oxford University Press, 2013, p. 4.
74. James Temple, "At This Rate, It's Going to Take Nearly 400 Years to Transform the Energy System," *MIT Technology Review*, 14 Mar 2018.
75. Mark Z. Jacobson et al., "100% Clean and Renewable Wind, Water, and Sunlight (WWS) All-sector Energy Roadmaps for the 50 United States," *Energy & Environmental Science*, vol. 8, no. 7, 2015.
76. Timothy Allen et al., "Radical Transitions from Fossil Fuel to Renewables," *Complex Systems and Social Practices in Energy Transitions*, Springer, 2017.
77. Peter J. Loftus et al., "A Critical Review of Global Decarbonization Scenarios," *Wiley Interdisciplinary Reviews: Climate Change*, vol. 6, no. 1, 2015.
78. Lewis C. King and Jeroen CJM van den Bergh, "Implications of Net Energy-return-on-investment for a Low-carbon Energy Transition," *Nature Energy*, vol. 3, 2018; Friends of the Earth Europe, "Sufficiency: Moving Beyond the Gospel of Eco-Efficiency," Mar 2018.
79. Based on World Bank data. See: Kris De Decker, "How Much Energy Do We Need?," *Demand*, 17 Jan 2018.
80. See *Fuel Poverty Action*, Energy Bill of Rights, nin.tl/EnergyVision
81. Kris De Decker, "How Much Energy Do We Need?," *Demand*, 17 Jan 2018; Simon Evans, "World Can Limit Global Warming to 1.5C 'Without BECCS,'" *Carbon Brief*, 13 Apr 2018.

82. René Kleijn, "The Energy Transition Consumes Large Amounts of Metals," *Green European Journal*, 11 Dec 2015.

83. Julie Cart, "The Power Compromise," *Los Angeles Times*, 5 Feb 2012.

84. Cited in Hamza Hamouchene, "Desertec: The Renewable Energy Grab?," *New Internationalist*, 1 Mar 2015.

85. Maria Kaika, "Our Sustainability Is Someone Else's Disaster," *Green European Journal*, 5 Jan 2018.

86. Sarah Krakoff, "Just Transitions?," *Law and Political Economy Blog*, 29 Jan 2018.

87. Laura Hamister, "Wind Development of Oaxaca, Mexico's Isthmus of Tehuantepec," *Mexican Law Review*, vol. 5, no. 1, 2012.

88. René Kleijn, "The Energy Transition Consumes Large Amounts of Metals," *Green European Journal*, 11 Dec 2015.

89. Cécile Bontron, "Rare-earth Mining in China Comes at a Heavy Cost for Local Villages," *Guardian*, 7 Aug 2012.

90. Simon Denyer, "Tibetans in Anguish as Chinese Mines Pollute Their Sacred Grasslands," *Washington Post*, 26 Dec 2016.

91. Andy Home, "Cobalt, the Heart of Darkness in the Shiny Electric Vehicle Story," *Reuters*, 28 Nov 2017.

92. Mark Dummett, "The Dark Side of Electric Cars," *Time*, 28 Sep 2017.

93. See: "A Transition Plan for Communities Affected by the Closings of Navajo Generating Station and Kayenta Mine," Institute for Energy Economics and DinéHózhó L3C, Jun 2017.

94. Dieter Helm, *Burn Out*, Yale University Press, 2017; Paul Stevens, "International Oil Companies," Chatham House, 5 May 2016; "Oil and Gas Industry Faces Significant Credit Risks from Carbon Transition," Moody's, 26 Apr 2017.

95. Sarah Kent, Bradley Olson and Georgi Kantchev, "Energy Companies Face Crude Reality," *Wall Street Journal*, 17 Feb 2017.

96. John A. Powell, Connie Cagampang Heller and Fayza Bundalli, "Systems Thinking and Race: Workshop Summary," California Endowment, 2011.

97. See: Alicia H. Puleo (ed), *Ecología y Género en diálogo interdisciplinar*, *Plaza y Valdés Editores*, 2015; Knut H. Sørensen, "Towards a Feminized Technology?" *Social Studies of Science*, vol. 22, no. 1, 1992; Junot Diaz, "Radical Hope Is Our Best Weapon," *On Being*, 14 Sep 2017.

98. On bodies and brutalities, see Begoña Dorronsoro, "El territorio cuerpo-tierra como espacio-tiempo de resistencias y luchas en las mujeres indígenas y originarias," *IV Coloquio Internacional de Doctorandos/as do CES*, 2013; Cabnal refers to territorial violence across a range of talks and articles. For a window into her vision, see: Lorena

Cabnal, "Acercamiento a la construcción de la propuesta de pensamiento epistémico de las mujeres indígenas feministas

99. Richard Wilkinson and Kate Pickett, *The Spirit Level*, Penguin, 2010.

100. Ibid; Danny Dorling, *The Equality Effect*, New Internationalist, 2017.

101. Eric Neumayer and Thomas Plümper, "The Gendered Nature of Natural Disasters," *Annals of the Association of American Geographers*, vol. 97, no. 3, 2007; Joshua Eastin, "Climate Change and Gender Equality in Developing States," *World Development*, vol. 107, 2018; Women and Gender Constituency, "Gender-Just Climate Solutions," 2016, nin.tl/GenderJustice

102. See the Climate Fair Shares resource: nin.tl/FairShares

103. Humberto Maturana Romesin and Gerda Verden-Zoller, "Biology of Love," *Focus Heilpaedagogik*, Ernst Reinhardt, 1996.

104. Jo Marchant, "Poorest Costa Ricans Live Longest," *Nature—News*, 3 Sep 2013; George Monbiot, "The Town That's Found a Potent Cure for Illness—Community," *Guardian*, 21 Feb 2018; Tony Juniper, *What Has Nature Ever Done for Us?*, Profile, 2013, p. 257.

105. Sarah Leonard and Nancy Fraser, "Capitalism's Crisis of Care," *Dissent*, Fall 2016.

106. Giorgos Kallis et al., "Friday Off," *Sustainability*, vol. 5, no. 4, 2013; Silvia Kega Ugalde, "Sumak kawsay, feminismos y post-crecimiento," *Postcrecimiento y Buen Vivir*, Friedrich-Ebert-Stiftung, 2014, pp. 355–374.

107. For example: Vatsal Chikani et al., "Vacations Improve Mental Health Among Rural Women," *WMJ-MADISON*, vol. 104, no. 6, 2005; Brooks B. Gump and Karen A. Matthews, "Are Vacations Good for Your Health?," *Psychosomatic Medicine*, vol. 62, no. 5, 2000; Kate Sparks et al., "The Effects of Hours of Work on Health," *Journal of Occupational and Organizational Psychology*, vol. 70, no. 4, 1997; Marianna Virtanen et al., "Long Working Hours and Cognitive Function," *American Journal of Epidemiology*, vol. 169, no. 5, 2009.

108. Giorgos Kallis et al., op cit.

109. David Graeber, "On the Phenomenon of Bullshit Jobs," *Strike! Magazine*, no. 3, Aug 2013.

110. See Robin Wall Kimmerer, *Braiding Sweetgrass*, Milkweed Editions, 2013.

111. Annie Leonard, *The Story of Stuff*, Constable, 2016, p. 3.

112. Paul Hawken, *Drawdown*, Penguin, 2017; Karl-Heinz Erb et al., "Unexpectedly Large Impact of Forest Management and Grazing on Global Vegetation Biomass," *Nature*, vol. 553, no. 7686, 2018; Bronson W. Griscom et al., "Natural Climate Solutions," *Proceedings of the National Academy of Sciences*, vol. 114, no. 44, 2017; Benjamin Quesada

et al., "Reduction of Monsoon Rainfall in Response to Past and Future Land Use and Land Cover Changes," *Geophysical Research Letters*, vol. 44, no. 2, 2017.

113. Michael Köhl, Prem R. Neupane and Neda Lotfiomran, "The Impact of Tree Age on Biomass Growth and Carbon Accumulation Capacity," *PLoS ONE*, vol. 12, no. 8, 2017.

114. Caleb Stevens et al., "Securing Rights, Combating Climate Change," *World Resources Institute*, July 2014.

115. RAISG et al., "Amazonian Indigenous Peoples Territories and Their Forests Related to Climate Change," Oct 2017, nin.tl/Amazon

116. Chris Arsenault, "How to Protect Peru's Rainforest?," *Reuters*, 3 Apr 2017.

117. Helena Durán, "El lado oscuro de la conservación," *DeJusticia*, 4 Sep 2017; John Vidal, "The Tribes Paying the Brutal Price of Conservation," *Observer*, 28 Aug 2016.

118. Stan Cox and Paul Cox, *How the World Breaks*, New Press, 2016, p. 104.

119. Ibid, p. 241; Richard Stone, "Cuba Embarks On a 100-year Plan to Protect Itself from Climate Change," *Science*, 10 Jan 2017.

120. Ahmedabad Municipal Corporation, "Ahmedabad Heat Action Plan 2015," 2015, nin.tl/Ahmedabad.

121. Jeroen Frank Warner, Arwin van Buuren and Jurian Edelenbos (eds), *Making Space for the River*, IWA Publishing, 2012.

122. Thomas J. Bassett and Charles Fogelman, "Déjà Vu or Something New?," *Geoforum*, vol. 48, Aug 2013; Mark Pelling, *Adaptation to Climate Change*, Routledge, 2010.

123. Eric Klinenberg, "Adaptation: How Can Cities Be 'Climate-proofed'?" *New Yorker*, 7 Jan 2013; Daniel P. Aldrich, *Building Resilience*, University of Chicago Press, 2012.

124. Jessica Ayers et al. (eds), *Community-Based Adaptation to Climate Change*, Routledge, 2014.

125. Daniel Macmillen Voskoboynik, "The Legal Limbo of Climate Refugees," *New Internationalist*, 14 Nov 2016; for more information, see the work of the Climate and Migration Coalition.

126. Sociologist Abdelmalek Sayad called it the "double absence," exclusion from home and host country.

127. Benjamin Glahn, "Climate Refugees?," *International Bar Association*, Jun 2009.

128. Gillian Judson and Rob Hopkins, "Gillian Judson: 'Education That Inspires Is What We're After,'" 26 Feb 2018, nin.tl/Gillian; Gill Helsby and Murray Saunders, "Taylorism, Tylerism and Performance Indicators," *Educational Studies*, vol. 19, no. 1, 1993.

129. Manish Jain and Shilpa Jain (eds), *Healing Ourselves from the Diploma Disease*, People's Institute for Rethinking Education and Development, 2005.
130. Yayo Herrera López, "El curriculum oculto antiecológico de los libros de texto, *Ambienta*, vol. 69, 2007.
131. Lant Pritchett, *The Rebirth of Education*, CGD Books, 2013.
132. Paulo Freire, *La educación como práctica de la libertad*, Siglo XXI Editores, 2007.
133. Gillian Judson, *Engaging Imagination in Ecological Education*, Practical Strategies for Teachers, University of British Columbia Press, 2017; Rachael Kessler, *The Soul of Education*, ASCD, 2000.
134. Stephen Cave and Sarah Darwin, "It's Not Easy Being Green," *Aeon*, 15 Mar 2016.
135. Robin Wall Kimmerer, op cit.
136. Tilly Alcanya, "To Prevent the Next Global Crisis, Don't Forget Today's Small Disasters," *The Conversation*, 12 Dec 2017.
137. Cited in Nicholas Thompson, "Our Minds Have Been Hijacked by Our Phones," *Wired*, 27 Jul 2017.
138. Arwa Mahdawi, "Can Cities Kick Ads?' *Guardian*, 12 Aug 2015.
139. Judy Atkinson, *Trauma Trails, Recreating Song Lines*, Spinifex Press, 2002.
140. David Bohm, *Wholeness and the Implicate Order*, Routledge, 1980, p. 3.
141. Ted Schettler, "Toward an Ecological View of Health," *Designing the 21st Century Hospital*, Center for Health Design and Health Care Without Harm, 2006.
142. Helen L. Berry et al., "The Case for Systems Thinking About Climate Change and Mental Health," *Nature Climate Change*, vol. 8, no. 4, 2018.
143. Physician John Howard has called for an "ecosystem health model." See: Michael Bentley, "An Ecological Public Health Approach to Understanding the Relationships Between Sustainable Urban Environments, Public Health and Social Equity," *Health Promotion International*, vol. 29, no. 3, 2013; A. Alonso Aguirre et al. (eds), *Conservation Medicine*, Oxford University Press, 2002.
144. The work of Richard Louv is emblematic in this area.
145. Fabio Caiazzo et al., "Air Pollution and Early Deaths in the United States," *Atmospheric Environment*, vol. 79, 2013.
146. Hanna Erixon Aalto, *Projecting Urban Natures*, Dent-de-Leone, 2017, p. 17.
147. Ana María Vásquez Duplat, "Feminismo y 'extractivismo urbano.'" *Nueva Sociedad*, no. 265, Sep-Oct 2016.

148. See: Randolph T. Hester, *Design for Ecological Democracy*, MIT Press, 2006; Charles Montgomery, *Happy City*, Macmillan, 2013.
149. Parinaz Motealleh, Maryam Zolfaghari and Mojtaba Parsaee, "Investigating Climate Responsive Solutions in Vernacular Architecture of Bushehr City," *HBRC Journal*, 2016.
150. Oriol Marquet and Carme Miralles-Guasch, "Walking Short Distances," *Transportation Research Part A: Policy and Practice*, vol. 70, 2014.

Chapter 14: What Then Must We Do?

1. Martin Lukacs, "Neoliberalism Has Conned Us into Fighting Climate Change as Individuals," *Guardian*, 17 Jul 2017.
2. Cited in Jack Heinemann, "The Seed Vault Flooding Is Only the Start of Our Problems," *Boston Review*, 26 May 2017.
3. Elif Shafak, "The Urgency of a Cosmopolitan Ideal as Nationalism Surges," *New Perspectives Quarterly*, vol. 31, no. 2, 2014.
4. See: Jean G. Boulton, Peter M. Allen and Cliff Bowman, *Embracing Complexity*, Oxford University Press, 2015.
5. Mark Engler and Paul Engler, *This Is an Uprising*, Nation Books, 2016, p. 282.
6. David Bollier, "A Just Transition and Progressive Philanthropy," 19 Apr 2016, nin.tl/Bollier
7. Mark Engler and Paul Engler, op cit, pp. xvii and 13; Paul Engler and Sophie Lasoff, *Resistance Guide*, Momentum, 2017.
8. Transnational Institute, "Women, Resistance and Counter-power," 2018, nin.tl/TNI
9. Erica Chenoweth and Maria J. Stephan, *Why Civil Resistance Works*, Columbia University Press, 2012.
10. See Paul Hawken, *Drawdown*, Penguin, 2017; Lars-Arvid Brischke et al., *Energy Sufficiency in Private Households Enabled by Adequate Appliances*, Wuppertal Institut für Klima, 2015.
11. See Kevin Anderson, cited in Terry Macalister, "Westerners Urged to Reduce Carbon Footprint," *Climate News Network*, 21 Jan 2017.
12. Ivetta Gerasimchuk et al., *Zombie Energy*, IISD, Feb 2017.
13. Martin Luther King Jr, "A Christmas Sermon," *The King Center*, 24 Dec 1967.
14. Jonathan Matthew Smucker, *Hegemony How-to*, AK Press, 2017.
15. Yehuda Amichai, *The Selected Poetry of Yehuda Amichai*, University of California Press, 2013, p. 34.
16. Joanna Macy and Molly Young Brown, *Coming Back to Life*, New Society, 1998.

17. Tom W. Smith, Jibun Kim and Jaesok Son, "Public Attitudes Toward Climate Change and Other Environmental Issues Across Countries," *International Journal of Sociology*, vol. 47, no. 1, 2017. See also the work of psychologist John Krosnick.
18. Ezra M. Markowitz and Azim F. Shariff, "Climate Change and Moral Judgment," *Nature and Climate Change*, vol. 2, no. 4, Apr 2012.
19. Matthew J. Hornsey and Kelly S. Fielding, "A Cautionary Note About Messages of Hope," *Global Environmental Change*, vol. 39, Jul 2016.
20. Arjun Appadurai, *The Future as Cultural Fact*, Verso, 2013.
21. Malini Ranganathan, "The Environment as Freedom," *Items*, 13 Jun 2017.
22. Khadija Ismayilova, "Azerbaijani Laundromat Shows How Regime Robs Its People to Feed Itself," *Guardian*, 5 Sep 2017.
23. India Bourke, "Here Are 16 Cities Tackling Inequality Through Climate Action Schemes," *CityMetric*, 28 Sep 2017.
24. Neema Pathak Broome and Ashish Kothari, "How an Ecuadorian Community Is Showing Its Government How to Really Live Well," *Radical Ecological Democracy*, 16 Dec 2017.
25. Truls Gulowsen, "In Norway, a Growing Movement Builds an Oil-Free Future," *The Leap Blog*, 27 Apr 2017.
26. Michal Nachmany et al., "Global Trends in Climate Change Legislation and Litigation," Grantham Institute, 2017.
27. Michelle Chen, "China's Green Movement," *In These Times*, 24 Feb 2014.
28. International Criminal Court—Office of the Prosecutor, "Policy Paper on Case Selection and Prioritisation," 15 Sep 2016.

Chapter 15: Hope, A Horizon

1. Naveeda Khan, "Dogs and Humans and What Earth Can Be," *Journal of Ethnographic Theory*, vol. 4, no. 3, 2014.
2. The biologist Daniel Pauly called this "shifting baseline syndrome," in Daniel Pauly, "Anecdotes and the Shifting Baseline Syndrome of Fisheries," *Trends in Ecology and Evolution*, vol. 10, no. 10, 1995.
3. Cited in Eduardo Galeano, *Las palabras andantes*, Catálogos, 2001, p. 230.
4. Mary Watkins and Helene Shulman, *Toward Psychologies of Liberation*, Palgrave Macmillan 2008, p. 111.
5. Oil Change International and Greenpeace, *Forecasting Failure*, Mar 2017.
6. Nick Buxton, "Defying Dystopia," *Roar Magazine*, no. 7, 2018.
7. Jeremy Adelman, "Why the Idea That the World Is in Terminal Decline Is So Dangerous," *Aeon*, 1 Nov 2017.

8. Mike Davis, *In Praise of Barbarians*, Haymarket Books, 2007, p. 259.

9. See the work of Donella Meadows; Timothy Allen et al., "Mapping Degrees of Complexity, Complicatedness, and Emergent Complexity," *Ecological Complexity*, 2017.

10. Vasily Grossman, *Life and Fate*, Vintage Books 2011, p. 394.

11. Howard Zinn, *You Can't Be Neutral on a Moving Train*, Beacon Press, 2002, p. 208.

12. Robin Wall Kimmerer, "The Intelligence in All Kinds of Life," *On Being*, 24 Feb 2016.

13. Wendell Berry writes, "People exploit what they have merely concluded to be of value, but they defend what they love, and to defend what we love we need a particularizing language, for we love what we particularly know." Cited in David Bollier, *Silent Theft*, Routledge 2003, p. 68.

14. Fernando Coronil, "The Future in Question," *Business as Usual*, New York University Press, 2011.

Index

About the Author

DANIEL MACMILLEN VOSKOBOYNIK is a journalist and activist with writing in Pacific Standard, Open Democracy, and New Internationalist. He co-founded and is co-editor of www.worldat1C.org, a communications initiative designed to humanize the ecological crisis and clarify its causes.
https://worldat1c.org/

ABOUT NEW SOCIETY PUBLISHERS

New Society Publishers is an activist, solutions-oriented publisher focused on publishing books for a world of change. Our books offer tips, tools, and insights from leading experts in sustainable building, homesteading, climate change, environment, conscientious commerce, renewable energy, and more—positive solutions for troubled times.

We're proud to hold to the highest environmental and social standards of any publisher in North America. This is why some of our books might cost a little more. We think it's worth it!

- We print all our books in North America, never overseas

- All our books are printed on **100% post-consumer recycled paper**, processed chlorine-free, with low-VOC vegetable-based inks (since 2002)

- Our corporate structure is an innovative employee shareholder agreement, so we're one-third employee-owned (since 2015)

- We're carbon-neutral (since 2006)

- We're certified as a B Corporation (since 2016)

At New Society Publishers, we care deeply about *what* we publish—but also about *how* we do business.

Download our catalog at https://newsociety.com/Our-Catalog or for a printed copy please email info@newsocietypub.com or call 1-800-567-6772 ext 111.

New Society Publishers

ENVIRONMENTAL BENEFITS STATEMENT

For every 5,000 books printed, New Society saves the following resources:[1]

33	Trees
3,020	Pounds of Solid Waste
3,322	Gallons of Water
4,334	Kilowatt Hours of Electricity
5,489	Pounds of Greenhouse Gases
24	Pounds of HAPs, VOCs, and AOX Combined
8	Cubic Yards of Landfill Space

[1] Environmental benefits are calculated based on research done by the Environmental Defense Fund and other members of the Paper Task Force who study the environmental impacts of the paper industry.